Standard Aircraft Handbook for Mechanics and Technicians

ABOUT THE EDITORS

Larry Reithmaier, a retired aerospace engineer, helped design and develop jet fighter and bomber aircraft and NASA spacecraft. He has a degree in mechanical engineering and holds the following FAA certificates: Commericial/ Instrument Pilot, Flight/Ground Instructor, and Airframe and Powerplant (A&P) Mechanic. The author of several technical books on aviation, Mr. Reithmaier has also written the *Aviation and Space Dictionary, Mach I and Beyond*, the *Private Pilot's Guide*, and the *Instrument Pilot's Guide*.

Ronald Sterkenburg started teaching at Purdue University in 1999 after a 20-year career in the U.S. Navy. He is the author or co-author of 10 books and more than 60 articles in national and international journals and conference proceedings. The focus of Dr. Sterkenburg's research is the repair of metallic and composite aircraft structures. He is a certificated Airframe and Powerplant (A&P) Mechanic, holds an Inspection Authorization (IA), and is a Designated Mechanic Examiner (DME).

Standard Aircraft Handbook for Mechanics and Technicians

Edited by Larry Reithmaier

and Ronald Sterkenburg

Seventh Edition

New York Chicago San Francisco Athens London
Madrid Mexico City Milan New Delhi
Singapore Sydney Toronto

11   LCR   20

ISBN 978-0-07-182679-2
MHID 0-07-182679-3

Sponsoring Editor
Larry S. Hager

Editing Supervisor
Stephen M. Smith

Production Supervisor
Richard C. Ruzycka

Acquisitions Coordinator
Bridget L. Thoreson

Project Manager
Yashmita Hota,
Cenveo® Publisher Services

Copy Editor
Megha Saini,
Cenveo Publisher Services

Proofreader
Barnali Ojha,
Cenveo Publisher Services

Indexer
Robert Swanson

Art Director, Cover
Jeff Weeks

Composition
Cenveo Publisher Services

Printed and bound by LSC Communications.

McGraw-Hill Education books are available at special quantity discounts to use as
premiums and sales promotions, or for use in corporate training programs. To contact a representative, please visit the Contact Us page at www.mhprofessional.com.

This book is printed on acid-free paper.

Contents

Preface

The *Standard Aircraft Handbook for Mechanics and Technicians* is presented in shop terms for the mechanics and technicians engaged in building, maintaining, overhauling, and repairing metal and composite aircraft. It is also useful for the student mechanic, who must acquire the basic mechanical skills fundamental to every technical specialty.

This handbook is a relatively complete guide to all basic shop practices, such as use of basic tools, drilling, riveting, sheet-metal forming, use of threaded fasteners, and installation of plumbing, cables, and electrical wiring. Chapters on nondestructive testing (NDT) and corrosion detection and control provide a guide to advanced technology inspection and detection equipment, techniques, and procedures.

For the Seventh Edition, the editors updated all existing chapters with new relevant information and figures to reflect current technologies, hardware, and materials used for aircraft maintenance. A new chapter was added to the handbook to provide information about composite materials such as carbon fiber used in modern aircraft models.

The information presented in this handbook was obtained from techniques and procedures developed by various aircraft and equipment manufacturers and is in general compliance with manufacturing specifications.

This handbook is not intended to replace, substitute for, or supersede any FAA regulations, shop and quality-control standards of an aircraft manufacturer, repair station, or manufacturer's maintenance manual.

Acknowledgments

Extensive use was made of data, information, illustrations, and photographs supplied by:

Lockheed Corporation
Rockwell International, NAAO
Federal Aviation Administration
Hi-Shear Corporation
Snap-On Tools
L.S. Starrett
The Aluminum Association
U.S. Industrial Tool & Supply
Lufkin Rule
Townsend Textron (Boots, Cherry fasteners)
Aeroquip
Centurion NDT Inc.
PRI Research and Development Corp.
LPS Laboratories Inc.
Parts suppliers catalogs
Boeing Aerospace Company
Airbus S.A.S.
Applied Composites Engineering (ACE)
Cirrus AIRCRAFT

DECIMAL EQUIVALENTS

Fraction	Decimal		Fraction	Decimal
1/64	.015625		33/64	.515625
1/32	.03125		17/32	.53125
3/64	.046875		35/64	.546875
1/16	.0625		9/16	.5625
5/64	.078125		37/64	.578125
3/32	.09375		19/32	.59375
7/64	.109375		39/64	.609375
1/8	.125		5/8	.625
9/64	.140625		41/64	.640625
5/32	.15625		21/32	.65625
11/64	.171875		43/64	.671875
3/16	.1875		11/16	.6875
13/64	.203125		45/64	.703125
7/32	.21875		23/32	.71875
15/64	.234375		47/64	.734375
1/4	.250		3/4	.750
17/64	.265625		49/64	.765625
9/32	.28125		25/32	.78125
19/64	.296875		51/64	.796875
5/18	.3125		13/16	.8125
21/64	.328125		53/64	.828125
11/32	.34375		27/32	.84375
23/64	.359375		55/64	.859375
3/8	.375		7/8	.875
25/64	.390625		57/64	.890625
13/32	.40625		29/32	.90625
27/64	.421875		59/64	.921875
7/16	.4375		15/16	.9375
29/64	.453125		61/64	.953125
15/32	.46875		31/32	.96875
31/64	.484375		63/64	.984375
1/2	.500		1	1.000

Introduction

Many aircraft configurations have been built, such as flying wing, tailless, canard, and biplane; however, the basic airplane configuration consists of a monoplane with a fuselage and tail assembly. See Figs. 1-1 and 1-2.

Although other construction methods are, or have been, used, such as wood, fabric, steel tube, composites, and plastics, the basic all-metal aluminum alloy structure predominates with steel and/or titanium in high-stress or high-temperature locations.

The airframe components are composed of various parts called *structural members* (i.e., stringers, longerons, ribs, formers, bulkheads, and skins). These components are joined by rivets, bolts, screws, and welding. Aircraft structural members are designed to carry a load or to resist stress. A single member of the structure could be subjected to a combination of stresses.

In designing an aircraft, every square inch of wing and fuselage, every rib, spar, and each metal fitting must be considered in relation to the physical characteristics of the metal of which it is made. Every part of the aircraft must be planned to carry the load to be imposed upon it. The determination of such loads is called *stress analysis*. Although planning the design is not the function of the aviation mechanic, it is nevertheless important that he or she understand and appreciate the stresses involved in order to avoid changes in the original design through improper repairs or poor workmanship.

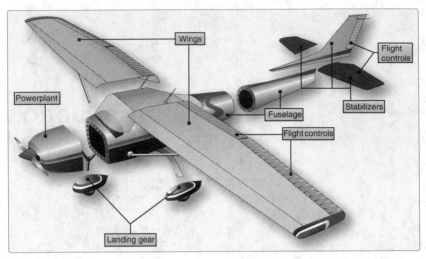

Figure 1-1   Major components of a piston-engine–powered light airplane.

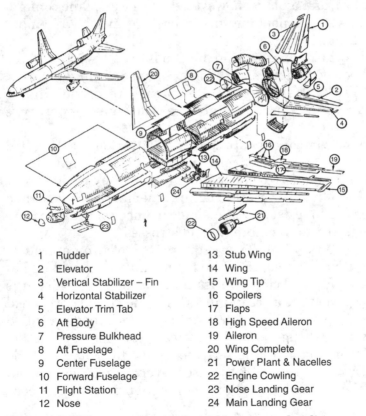

1	Rudder	13	Stub Wing
2	Elevator	14	Wing
3	Vertical Stabilizer – Fin	15	Wing Tip
4	Horizontal Stabilizer	16	Spoilers
5	Elevator Trim Tab	17	Flaps
6	Aft Body	18	High Speed Aileron
7	Pressure Bulkhead	19	Aileron
8	Aft Fuselage	20	Wing Complete
9	Center Fuselage	21	Power Plant & Nacelles
10	Forward Fuselage	22	Engine Cowling
11	Flight Station	23	Nose Landing Gear
12	Nose	24	Main Landing Gear

Figure 1-2   Major components of a turbine-powered airliner.

Fuselage Structure

The monocoque (single shell) fuselage relies largely on the strength of the skin or covering to carry the primary stresses. Most aircraft, however, use the semimonocoque design inasmuch as the monocoque type does not easily accommodate concentrated load points, such as landing gear fittings, powerplant attachment, wing fittings, etc.

The semimonocoque fuselage (Fig. 1-3) is constructed primarily of aluminum alloy, although steel and titanium are used in areas of high temperatures and/or high stress. Primary loads are taken by the longerons, which usually extend across several points of support. The longerons are supplemented by other longitudinal members, called *stringers*. Stringers are more numerous and lighter in weight than longerons and usually act as stiffeners. The vertical structural members are referred to as *bulkheads*, *frames*, and *formers*. The heaviest of these vertical members are located at intervals to carry concentrated loads and at points where fittings are used to attach other units, such as the wings, powerplants, and stabilizers.

Location Numbering Systems

Various numbering systems are used to facilitate the location of specific wing frames, fuselage bulkheads, or any other structural

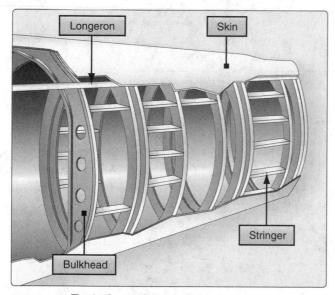

Figure 1-3   Typical metal aircraft fuselage structure.

members on an aircraft. Most manufacturers use some system of station marking; for example, the nose of the aircraft may be designated zero station, and all other stations are located at measured distances in inches behind the zero station. Thus, when a blueprint reads "fuselage frame station 137," that particular frame station can be located 137 inches behind the nose of the aircraft. However, the zero station may not be the nose of the fuselage, as in Fig. 1-4.

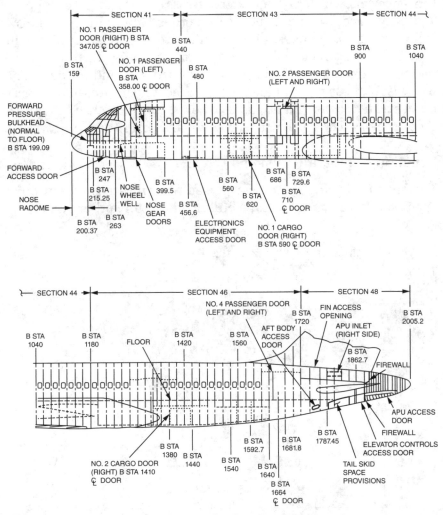

Figure 1-4 Typical drawing showing fuselage stations.

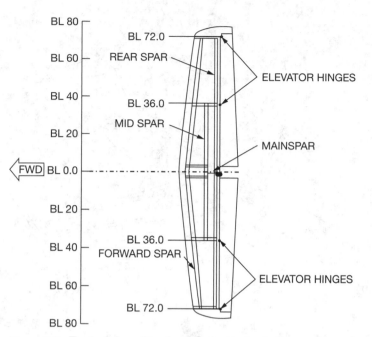

Figure 1-5 Typical drawing showing buttock stations.

To locate structures to the right or left of the center line of an aircraft, many manufacturers consider the center line as a zero station for structural member location to its right or left as shown in Fig. 1-5. With such a system, the stabilizer frames can be designated as being so many inches right or left of the aircraft center line.

1. Fuselage stations (F.S.) are numbered in inches from a reference or zero point known as the *reference datum.* The reference datum is an imaginary vertical plane at or near the nose of the aircraft from which all horizontal distances are measured. The distance to a given point is measured in inches parallel to a center line extending through the aircraft from the nose through the center of the tail cone.

2. Buttock line or butt line (B.L.) is a width measurement left or right of, and parallel to, the vertical center line.

3. Water line (W.L.) is the measurement of height in inches perpendicular from a horizontal plane located a fixed number of inches below the bottom of the aircraft fuselage; see Fig. 1-6.

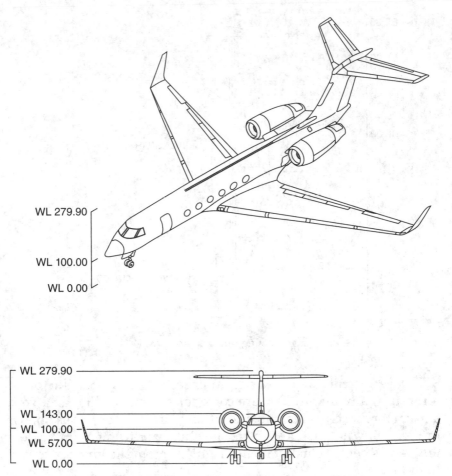

Figure 1-6   Typical drawing showing waterline station numbers.

Chapter 10, Aircraft Drawings, provides additional information regarding aircraft drawings generally referred to as *blueprints*.

Wing Structure

The wings of most aircraft are of cantilever design; that is, they are built so that no external bracing is needed. The skin is part of the wing structure and carries part of the wing stresses. Other aircraft wings use external bracings (struts) to assist in supporting the wing and carrying the aerodynamic and landing loads. Aluminum alloy is primarily used in wing construction.

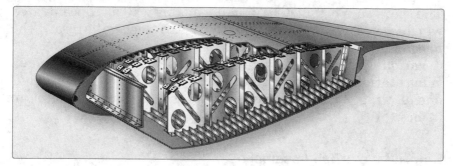

Figure 1-7 All-metal wing structure.

The internal structure is made up of spars and stringers running spanwise, and ribs and formers running chordwise (leading edge to trailing edge). See Fig. 1-7. The spars are the principal structural members of the wing. The skin is attached to the internal multiengine members and can carry part of the wing stresses. During flight, applied loads, which are imposed on the wing structure, are primarily on the skin. From the skin, they are transmitted to the ribs and from the ribs to the spars.

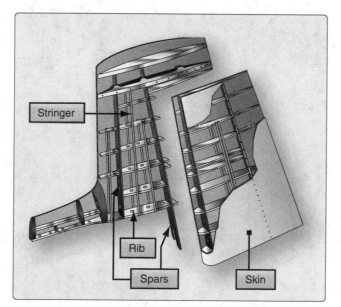

Figure 1-8 Typical vertical stabilizer and rudder construction.

The spars support all distributed loads, as well as concentrated weights, such as fuselage, landing gear, and, on aircraft, the nacelles or pylons.

Various points on the wing are located by station number. Wing station 0 (zero) is located at the center line of the fuselage, and all wing stations are measured outboard from that point, in inches.

Empennage or Tail Assembly

The fixed and movable surfaces of the typical tail assembly (Fig. 1-8) are constructed similarly to the wing. Each structural member absorbs some of the stress and passes the remainder to other members and, eventually, to the fuselage.

2

Tools and How to Use Them

Safety Considerations

Before commencing work on an aircraft, personal safety must become habit. Putting on safety glasses must be as much a part of the act of drilling a hole as picking up the drill motor.

The responsibility for this attitude lies with the mechanic, but this responsibility goes further. A mechanic's family needs him whole, with both eyes intact, both hands with all fingers intact, and above all, in good health.

Safety glasses or face shields must be worn during all of the following operations:

- Drilling
- Reaming
- Countersinking
- Driving rivets
- Bucking rivets
- Operating rivet squeezer
- Operating any power tool
- Near flying chips or around moving machinery

Ear plugs should be used as protection against the harsh noises of the rivet gun and general factory din. If higher noise levels than the rivet gun are experienced, a full-ear-coverage earmuff should be used because it is a highly sound-absorbent device.

For people with long hair, a snood-type cap that keeps the hair from entangling with turning drills should be worn. Shirt sleeves should be short and long sleeves should be rolled up at least to the elbow. Closed-toe, low-heel shoes should be worn. Open-toed shoes, sandals, ballet slippers, moccasins, and canvas-type shoes offer little or no protection for feet and should not be worn in the shop or factory. Safety shoes are recommended.

Compressed air should not be used to clean clothes or equipment.

General-Purpose Hand Tools

Hammers

Hammers include ball-peen and soft hammers (Fig. 2-1). The ball-peen hammer is used with a punch, with a chisel, or as a peening (bending, indenting, or cutting) tool. Where there is danger of scratching or marring the work, a soft hammer (for example, brass, plastic, or rubber) is used. Most accidents with hammers occur when the hammerhead loosens. The hammer handle must fit the head tightly. A sweaty palm or an oily or greasy handle might let the hammer slip. Oil or grease on the hammer face might cause the head to slip off the work and cause a painful bruise. Striking a hardened steel surface sharply with a ball-peen hammer is a safety hazard. Small pieces of sharp, hardened steel might break from the hammer and also break from the hardened steel. The result might be an eye injury or damage to the work or the hammer. An appropriate soft hammer should be used to strike hardened steel. If a soft hammer is not available, a piece of copper, brass, fiber, or wood material should be placed on the hardened steel and struck with the hammer, not the hardened steel.

Ball peen   Straight peen   Cross peen   Tinner's mallet   Riveting hammer

Figure 2-1   Types of hammers.

Screwdrivers

The screwdriver is a tool for driving or removing screws. Frequently used screwdrivers include the common, crosspoint, and offset. Also in use are various screwdriver bits that are designed to fit screws with special heads. These special screwdrivers are covered in Chap. 6.

A common screwdriver must fill at least 75 percent of the screw slot (Fig. 2-2). If the screwdriver is the wrong size, it will cut and burr the screw slot, making it worthless. A screwdriver with the wrong blade size might slip and damage adjacent parts of the structures. The common screwdriver is used only where slotted head screws or fasteners are used on aircraft.

The two common recessed head screws are the Phillips and the Reed and Prince. As shown in Fig. 2-2, the Reed and Prince recessed head forms a perfect cross. The screwdriver used with this screw is pointed on the end. Because the Phillips screw has a slightly larger center in the cross, the Phillips screwdriver is blunt on the end. The Phillips screwdriver is not interchangeable with the Reed and Prince. The use of the wrong type of screwdriver results in mutilation of the screwdriver and the screwhead. A screwdriver should not be used for chiseling or prying.

Pliers

The most frequently used pliers in aircraft repair work include the slip-joint, longnose, diagonal-cutting, water-pump, and vise-grip types as shown in Fig. 2-3. The size of pliers indicates their overall length, usually ranging from 5 to 12 inches. In repair work, 6-inch, slip-joint pliers are the preferred size.

Slip-joint pliers are used to grip flat or round stock and to bend small pieces of metal to desired shapes. Long-nose pliers are used to reach where the fingers alone cannot and to bend small pieces of metal. Diagonal-cutting pliers or diagonals or dikes are used to perform such work as cutting safety wire and removing cotter pins. Water-pump pliers, which have extra-long handles, are used to obtain a very powerful grip. Vise-grip pliers (sometimes referred to as a *vise-grip wrench*) have many uses. Examples are to hold small work as a portable vise, to remove broken studs, and to pull cotter pins.

Phillips Screwdriver

Phillips   Reed & Prince

Offset Screwdriver

Flat-blade screwdriver

Figure 2-2 Types of screwdrivers.

Pliers are not an all-purpose tool. They are not to be used as a wrench for tightening a nut, for example. Tightening a nut with pliers causes damage to both the nut and the plier jaw serrations. Also, pliers should not be used as a prybar or as a hammer.

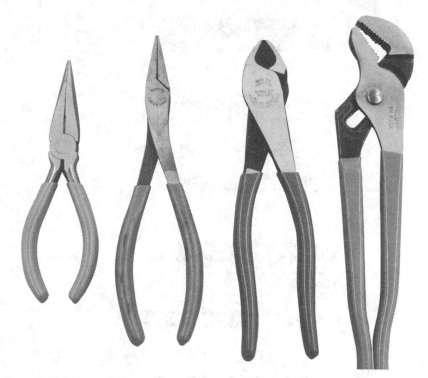

Figure 2-3 Types of pliers (from left to right: needle-nose, duckbill, diagonal cutter, and water-pump pliers).

Punches

Punches are used to start holes for drilling; to punch holes in sheet metal; to remove damaged rivets, pins, or bolts; and to align two or more parts for bolting together. A punch with a mush-roomed head should never be used. Flying pieces might cause an injury. Typical punches used by the aircraft mechanic are shown in Fig. 2-4.

Wrenches

Wrenches are tools used to tighten or remove nuts and bolts. The wrenches that are most often used are shown in Fig. 2-5a: open-end, box-end, adjustable, socket, and Allen wrenches. All have special advantages. The good mechanic will choose the one best suited for the job at hand. Sockets are used with the various

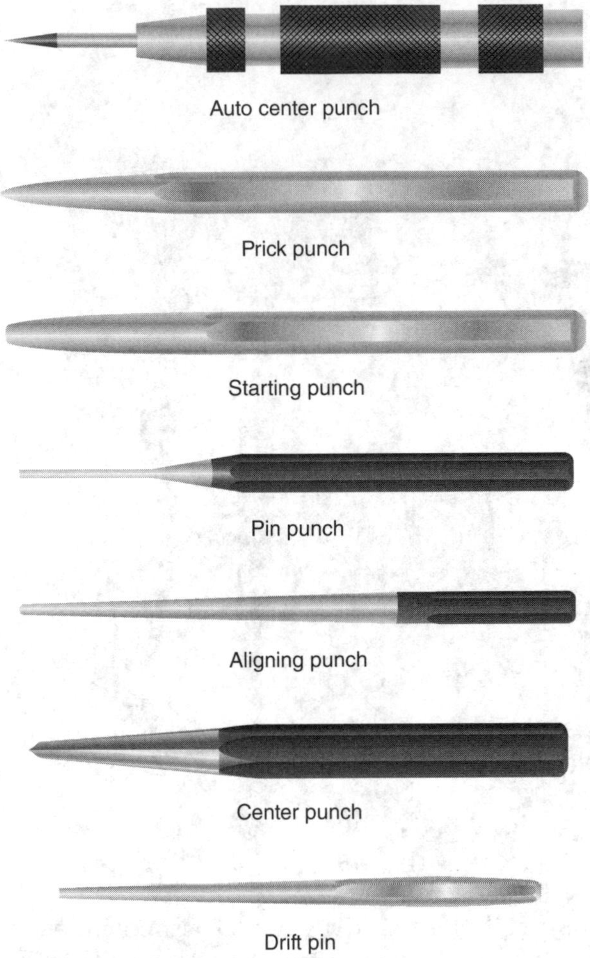

Auto center punch

Prick punch

Starting punch

Pin punch

Aligning punch

Center punch

Drift pin

Figure 2-4 Typical punches.

handles (ratchet, hinge, and speed) and extension bars are shown in Fig. 2-5b. Extension bars come in various lengths. The ratchet handle and speed wrench can be used in conjunction with suitable adapters and various type screwdriver bits to quickly install or remove special-type screws. However, if screws must be torqued to a specific torque value, a torque wrench must be used. Adjustable wrenches should be used only when other wrenches do not fit. To prevent rounding off the corners of a nut, properly adjust the wrench. The wrench should always be pulled so that the handle moves toward the adjustable jaw. A wrench should always be pulled. It is dangerous to push on it. A pipe should not be used to increase wrench leverage. Doing so might break the wrench. A wrench should never be used as a hammer.

Proper torquing of nuts and bolts is important. Overtorquing or undertorquing might set up a hazardous condition. Specified torque values and procedures should always be observed.

Figure 2-5a   Wrenches and sockets (from top to bottom: ratchet wrench, open-end ratchet combination wrench, flare-nut wrench, box-end wrench, open-ended wrench, and combination wrench).

Figure 2-5b   Socket set (sockets, extension, ratchet, universal joint, flare-nut, and open-ended extension).

Torque wrenches.   The three most commonly used torque wrenches are the flexible beam, rigid, and ratchet types (Fig. 2-6). New electronic setting-type torque wrenches are now available that provide a high accuracy. When using the flexible-beam and rigid-frame torque wrenches, the torque value is read visually on a dial or scale

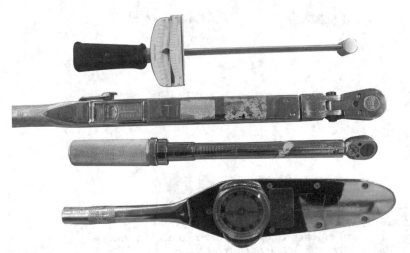

Figure 2-6   Common types of torque wrenches.

mounted on the handle of the wrench. To ensure that the amount of torque on the fasteners is correct, all torque wrenches must be tested at least once per month (or more often, if necessary).

The standard torque table presented in Chap. 6 should be used as a guide in tightening nuts, studs, bolts, and screws whenever specific torque values are not called out in maintenance procedures.

Metal-Cutting Tools

Hand snips

Hand snips serve various purposes. Straight, curved, hawksbill, and aviation snips are commonly used (Fig. 2-7). Straight snips are used to cut straight lines when the distance is not great enough to use a squaring shear, and to cut the outside of a curve. The other types are used to cut the inside of curves or radii. Snips should never be used to cut heavy sheet metal.

Aviation snips are designed especially to cut heat-treated aluminum alloy and stainless steel. They are also adaptable for enlarging small holes. The blades have small teeth on the cutting edges and are shaped to cut very small circles and irregular outlines. The handles are the compound-leverage type, making it

Figure 2-7   Various types of snips.

possible to cut material as thick as 0.051 inch. Aviation snips are available in three types, those that cut straight, those that cut from right to left, and those that cut from left to right.

Unlike the hacksaw, snips do not remove any material when the cut is made, but minute fractures often occur along the cut. Therefore, cuts should be made about ½₂ inch from the layout line and finished by hand-filing down to the line.

Hacksaws

The common hacksaw has a blade, a frame, and a handle. The handle can be obtained in two styles: pistol grip and straight. A pistol-grip hacksaw is shown in Fig. 2-8. When installing a blade in a hacksaw frame, the blade should be mounted with the teeth pointing forward, away from the handle.

Blades are made of high-grade tool steel or tungsten steel and are available in sizes from 6 to 16 inches in length. The 10-inch blade is most commonly used. The two types include the all-hard blade and the flexible blade. In flexible blades, only the teeth are hardened. Selection of the best blade for the job involves finding the right type of pitch. An all-hard blade is best for sawing brass, tool steel, cast iron, and heavy cross-section materials. A flexible blade is usually best for sawing hollow shapes and metals having a thin cross section.

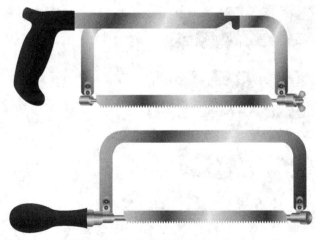

Figure 2-8 Pistol and straight grip hacksaws.

A. Mild Materials In Large Sections	B. Harder Materials In Large Sections	C. Unusual Work Shapes	D. Pipes, Tubing, Conduit
Choose coarse tooth blade to provide plenty of chip clearance, faster cutting.	Choose blade with finer teeth than in A to distribute cutting load over more teeth while still maintaining good chip clearing action.	Choose blade to always keep two or more teeth in contact with narrowest section. Coarse tooth blades straddle work, strip out teeth.	Choose blade with finest teeth per inch to keep two or more teeth in contact with wall. Keep inside of work free of chip accumulation.
Hand Blades— 14 Teeth per Inch	Hand Blades— 18 Teeth per Inch	Hand Blades— 24 Teeth per Inch	Hand Blades— 32 Teeth per Inch
Power Blades— 4 to 6 Teeth per Inch	Power Blades— 6 to 10 Teeth per Inch	Power Blades— 10 to 14 Teeth per Inch	Power Blades— 14 Teeth per Inch

Figure 2-9 Typical uses for various pitch hacksaw blades.

The pitch of a blade indicates the number of teeth per inch. Pitches of 14, 18, 24, and 32 teeth per inch are available. See Fig. 2-9.

1. Select an appropriate saw blade for the job.

2. Assemble the blade in the frame so that the cutting edge of the teeth points away from the handle.

3. Adjust tension of the blade in the frame to prevent the saw from buckling and drifting.

4. Clamp the work in the vise in such a way that it will provide as much bearing surface as possible and will engage the greatest number of teeth.

5. Indicate the starting point by nicking the surface with the edge of a file to break any sharp corner that might strip the teeth. This mark will also aid in starting the saw at the proper place.

6. Hold the saw at an angle that will keep at least two teeth in contact with the work at all times. Start the cut with a light, steady, forward stroke just outside the cutting line. At the end of the stroke, relieve the pressure and draw the blade back. (The cut is made on the forward stroke.)

7. After the first few strokes, make each stroke as long as the hacksaw frame will allow. This will prevent the blade from overheating. Apply just enough pressure on the forward stroke to cause each tooth to remove a small amount of metal. The strokes should be long and steady with a speed not more than 40 to 50 strokes per minute.

8. After completing the cut, remove chips from the blade, loosen tension on the blade, and return the hacksaw to its proper place.

Chisels

A chisel is a hard steel cutting tool that can be used to cut and chip any metal softer than the chisel itself. It can be used in restricted areas and for such work as shearing rivets, or splitting seized or damaged nuts from bolts (Fig. 2-10).

The size of a flat cold chisel is determined by the width of the cutting edge. Lengths will vary, but chisels are seldom fewer than 5 inches or more than 8 inches long.

A chisel should be held firmly in one hand. With the other hand, the chisel head should be struck squarely with a ball-peen hammer.

When cutting square corners or slots, a special cold chisel, called a *cape chisel*, should be used. It is like a flat chisel, except that the cutting edge is very narrow. It has the same cutting angle and is held and used in the same manner as any other chisel.

Rounded or semicircular grooves and corners that have fillets should be cut with a roundnose chisel. This chisel is also used to recenter a drill that has moved away from its intended center.

The diamond-point chisel is tapered square at the cutting end, then ground at an angle to provide the sharp diamond point. It is used to cut or for cutting grooves and inside sharp angles.

Files

Files are used to square ends, file rounded corners, remove burrs and slivers from metal, straighten uneven edges, file holes and slots, and smooth rough edges. Common files are shown in Fig. 2-11.

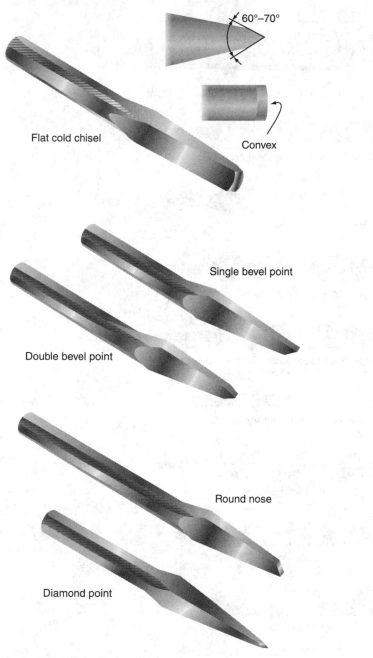

60°–70°

Flat cold chisel

Convex

Single bevel point

Double bevel point

Round nose

Diamond point

Figure 2-10   Chisels.

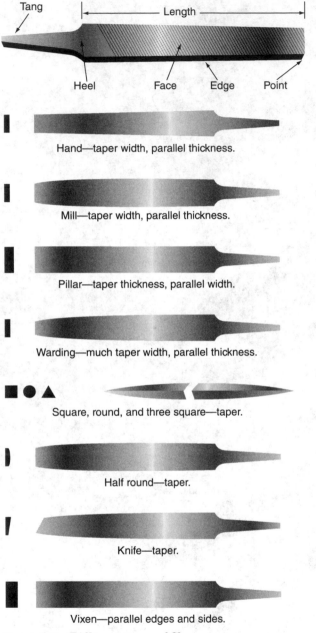

Figure 2-11   Different types of files.

Files are usually made in two styles: single cut and double cut. The single-cut file has a single row of teeth extending across the face at an angle of 65 to 85 degrees with the length of the file. The size of the cuts depends on the coarseness of the file. The double-cut file has two rows of teeth that cross each other. For general work, the angle of the first row is 40 to 45 degrees. The first row is generally referred to as *overcut*; the second row is called *upcut*. The upcut is somewhat finer and not so deep as the overcut.

The following methods are recommended for using files:

- *Crossfiling* Before attempting to use a file, place a handle on the tang of the file. This is essential for proper guiding and safe use. In moving the file endwise across the work (commonly known as *crossfiling*), grasp the handle so that its end fits into and against the fleshy part of your palm with your thumb lying along the top of the handle in a lengthwise direction. Grasp the end of the file between your thumb and first two fingers. To prevent undue wear, relieve the pressure during the return stroke.

- *Drawfiling* A file is sometimes used by grasping it at each end, crosswise to the work, then moving it lengthwise with the work. When done properly, work can be finished somewhat finer than when crossfiling with the same file. In drawfiling, the teeth of the file produce a shearing effect. To accomplish this shearing effect, the angle at which the file is held, with respect to its line of movement, varies with different files, depending on the angle at which the teeth are cut. Pressure should be relieved during the backstroke.

- *Rounding corners* The method used in filing a rounded surface depends upon its width and the radius of the rounded surface. If the surface is narrow or if only a portion of a surface is to be rounded, start the forward stroke of the file with the point of the file inclined downward at approximately a 45-degree angle. Using a rocking-chair motion, finish the stroke with the heel of the file near the curved surfaced. This method allows use of the full length of the file.

- *Removing burred or slivered edges* Practically every cutting operation on sheet metal produces burrs or slivers. These must be removed to avoid personal injury and to prevent scratching and marring of parts to be assembled. Burrs and slivers will prevent parts from fitting properly and should always be removed from the work as a matter of habit.

Particles of metal collect between the teeth of a file and might make deep scratches in the material being filed. When these particles of metal are lodged too firmly between the teeth and cannot be removed by tapping the edge of the file, remove them with a file card or wire brush. Draw the brush across the file so that the bristles pass down the gullet between the teeth.

Drilling and countersinking

Drilling and countersinking techniques are covered in Chap. 4.

Reamers

Reamers and reaming technique are covered in Chap. 4.

Layout and Measuring Tools

Layout and measuring devices are precision tools. They are carefully machined, accurately marked, and, in many cases, consist of very delicate parts. When using these tools, be careful not to drop, bend, or scratch them. The finished product will be no more accurate than the measurements or the layout; therefore, it is very important to understand how to read, use, and care for these tools.

Rules

Rules are made of steel and are either rigid or flexible. The flexible steel rule will bend, but it should not be bent intentionally because it could be broken rather easily (Fig. 2-12).

In aircraft work, the unit of measure most commonly used is the inch. The inch is separated into smaller parts by means of either common or decimal fraction divisions. The fractional divisions for an inch are found by dividing the inch into equal parts: halves ($\frac{1}{2}$), quarters ($\frac{1}{4}$), eighths ($\frac{1}{8}$), sixteenths ($\frac{1}{16}$), thirty-seconds ($\frac{1}{32}$), and sixty-fourths ($\frac{1}{64}$). The fractions of an inch can be expressed in decimals called decimal equivalents of an inch.

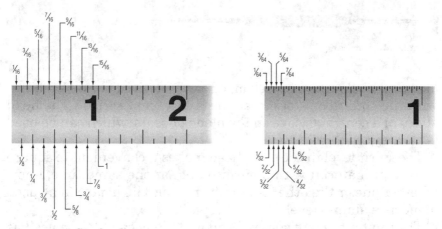

Figure 2-12 Steel rules.

For example, ⅛ inch is expressed as 0.125 (one hundred twenty-five ten-thousandths of an inch), or more commonly, twelve and one-half thousandths (see decimal equivalents chart on page xvi).

Rules are manufactured with two presentations: divided or marked in common fractions; divided or marked in decimals or divisions of 0.01 inch. A rule can be used either as a measuring tool or as a straightedge.

Combination sets

The combination set (Fig. 2-13), as its name implies, is a tool with several uses. It can be used for the same purposes as an ordinary trisquare, but it differs from the trisquare in that the head slides

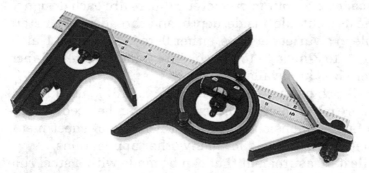

Figure 2-13 A combination set.

Figure 2-14 Scriber.

along the blade and can be clamped at any desired place. Combined with the square or stock head are a level and scriber. The head slides in a central groove on the blade or scale, which can be used separately as a rule.

The spirit level in the stock head makes it convenient to square a piece of material with a surface and, at the same time, know whether one or the other is plumb or level. The head can be used alone as a simple level.

The combination of square head and blade can also be used as a marking gauge (to scribe at a 45-degree angle), as a depth gauge, or as a height gauge.

Scriber

The scriber (Fig. 2-14) is used to scribe or mark lines on metal surfaces.

Dividers and calipers

Dividers have two legs tapered to a needle point and joined at the tip by a pivot. They are used to scribe circles and to transfer measurements from the rule to the work.

Calipers are used to measure diameters and distances or to compare distances and sizes. The most common types of calipers are the inside and the outside calipers (see Fig. 2-15).

Micrometer calipers.   Four micrometer calipers are each designed for a specific use: outside, inside, depth, and thread. Micrometers are available in a variety of sizes, either 0- to ½-inch, 0- to 1-inch, 1- to 2-inch, 2- to 3-inch, 3- to 4-inch, 4- to 5-inch, or 5- to 6-inch sizes. Larger sizes are available.

The 0- to 1-inch outside micrometer (Fig. 2-16) is used by the mechanic more often than any other type. It can be used to measure the outside dimensions of shafts, thickness of sheet metal stock, diameter of drills, and for many other applications.

The smallest measurement that can be made with a steel rule is one sixty-fourth of an inch in common fractions, and one

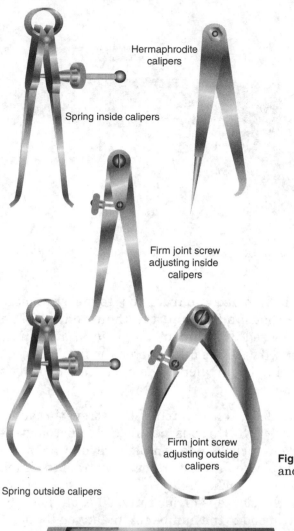

Hermaphrodite
calipers

Spring inside calipers

Firm joint screw
adjusting inside
calipers

Firm joint screw
adjusting outside
calipers

Figure 2-15 Typical outside and inside calipers.

Spring outside calipers

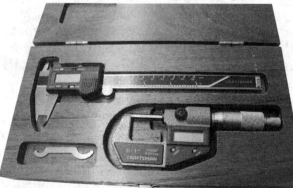

Figure 2-16 Digital micrometer and caliper.

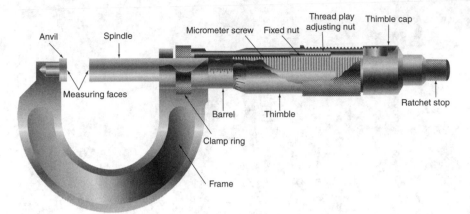

Figure 2-17 Parts of a micrometer.

one-hundredth of an inch in decimal fractions. To measure more closely than this (in thousandths and ten-thousandths of an inch), a micrometer is used. If a dimension given in a common fraction is to be measured with the micrometer, the fraction must be converted to its decimal equivalent. The micrometer consists of several parts as shown in Fig. 2-17.

Reading a micrometer. Because the pitch of the screw thread on the spindle is $\frac{1}{40}$ inch (or 40 threads per inch in micrometers graduated to measure in inches), one complete revolution of the thimble advances the spindle face toward or away from the anvil face precisely $\frac{1}{40}$ inch, 0.025 inch.

The reading line on the sleeve is divided into 40 equal parts by vertical lines that correspond to the number of threads on the spindle. Therefore, each vertical line designates $\frac{1}{40}$ inch or 0.025 inch, and every fourth line, which is longer than the others, designates hundreds of thousandths. For example: the line marked "1" represents 0.100 inch, the line marked "2" represents 0.200 inch, and the line marked "3" represents 0.300 inch, etc.

The beveled edge of the thimble is divided into 25 equal parts with each line representing 0.001 inch and every line numbered consecutively. Rotating the thimble from one of these lines to the next moves the spindle longitudinally $\frac{1}{2}$ of 0.025 inch, or 0.001 inch; rotating two divisions represents 0.002 inch, etc. Twenty-five divisions indicate a complete revolution, 0.025 inch or $\frac{1}{40}$ of an inch.

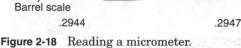

.2944 .2947

Figure 2-18 Reading a micrometer.

To read the micrometer in thousandths, multiply the number of vertical divisions visible on the sleeve by 0.025 inch; to this add, the number of thousandths indicated by the line on the thimble that coincides with the reading line on the sleeve.

Example: Refer to Fig. 2-18.

Some micrometers are equipped with a vernier scale that makes it possible to directly read the fraction of a division that is indicated on the thimble scale. The vernier graduations divide the 0.001 inch graduation on the thimble into 10 equal parts, each equal to 0.0001 inch. Newer digital micrometers as shown in Fig. 2-16 are very easy to read on a LCD screen and provide accurate measurements to 0.0001 inch.

Example: Refer to Fig. 2-18. In the first example in Fig. 2-18, the barrel reads 0.275 inch and the thimble reads more than 0.019 inch. The number 1 graduation on the thimble is aligned exactly with the number 4 graduation on the vernier scale. Thus, the final reading is 0.2944 inch. In the second example in Fig. 2-18, the barrel reads 0.275 inch, and the thimble reads more than 0.019 inch and less than 0.020 inch. On the vernier scale the number 7 graduation coincides with a line on the thimble. This means that the thimble reading would be 0.0197 inch. Adding this to the barrel reading of 0.275 inch gives a total measurement of 0.2947 inch.

Slide calipers

Slide calipers are used to measure the length of an object. This versatile tool can measure inside, outside, and depth dimensions. Vernier, dial, and digital calipers are available as shown in Fig. 2-19.

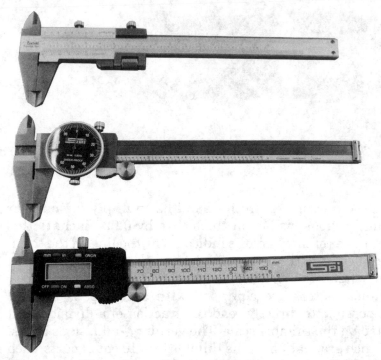

Figure 2-19 Slide calipers (from top to bottom: vernier, dial, and digital calipers).

Taps and Dies

A tap is used to cut threads on the inside of a hole and a die is to cut external threads on round stock. Taps and dies are made of hard-tempered steel and ground to an exact size. Four threads can be cut with standard taps and dies: national coarse, national fine, national extra fine, and national pipe.

Hand taps are usually provided in sets of three taps for each diameter and thread series. Each set contains a taper, a plug, and a bottoming tap. The taps in a set are identical in diameter and cross section; the only difference is the amount of taper (Fig. 2-20).

The taper tap is used to begin the tapping process because it is tapered back for six to seven threads. This tap cuts a complete thread when it is needed to tap holes that extend through thin sections. The plug tap supplements the taper tap for tapping holes in thick stock.

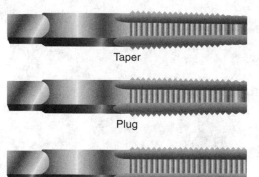

Taper

Plug

Bottoming

Figure 2-20 Hand taps.

Adjusting screw

(a) Adjustable round split die

Figure 2-21 Die types.

(b) Plain round split die

The bottoming tap is not tapered. It is used to cut full threads to the bottom of a blind hole.

Dies can be classified as adjustable round split and plain round split (Fig. 2-21). The adjustable-split die has an adjusting screw that can be controlled. Solid dies are not adjustable; therefore, several thread fits cannot be cut.

Many wrenches turn taps and dies: T-handle, adjustable tap, and diestock for round split dies (Fig. 2-22) are common. Information on thread sizes, fits, types, and the like, is in Chap. 6.

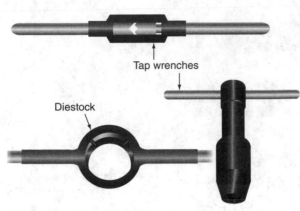

Figure 2-22   Diestock and tap wrenches.

Shop Equipment

Only the simpler metalworking machines, such as used in the service field, are presented in this manual. These include the powered and nonpowered metal-cutting machines, such as the various types of saws, powered and nonpowered shears, and nibblers. Also included is forming equipment (both power-driven and nonpowered), such as brakes and forming rolls, the bar folder, and shrinking and stretching machines. Factory equipment, such as hydropresses, drop-forge machines, and sparmills, for example, are not described.

Holding devices

Vises and clamps are used to hold materials of various kinds on which some type of operation is being performed. The operation and the material that is held determines which holding device is used. A typical vise is shown in Fig. 2-23.

Figure 2-23   A machinist's vise.

Figure 2-24 Throatless shears (Beverly shears).

Squaring shears

Squaring shears provide a convenient means of cutting and squaring metal. Three distinctly different operations can be performed on the squaring shears:

- Cutting to a line
- Squaring
- Multiple cutting to a specific size

A squaring shear is shown in Chap. 3.

Throatless shears

Throatless shears (Fig. 2-24) are best used to cut 10-gauge mild carbon steel sheet metal and 12-gauge stainless steel. The shear gets its name from its construction; it actually has no throat. It has no obstructions during cutting because the frame is throatless. A sheet of any length can be cut, and the metal can be turned in any direction to cut irregular shapes. The cutting blade (top blade) is operated by a hand lever.

Bar folder

The bar folder (Fig. 2-25) is designed to make bends or folds along edges of sheets. This machine is best suited for folding small hems,

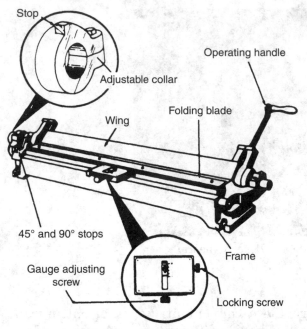

Figure 2-25 Manually operated bar folder.

flanges, seams, and edges to be wired. Most bar folders have a capacity for metal up to 22 gauge thickness and 42 inches long.

Sheet-metal brake

The sheet-metal brake (Fig. 2-26) has a much greater range of usefulness than the bar folder. Any bend formed on a bar folder can be made on the sheet-metal brake. The bar folder can form a bend or edge only as wide as the depth of the jaws. In comparison, the sheet-metal brake allows the sheet that is to be folded or formed to pass through the jaws from front to rear without obstruction.

Slip roll former

The slip roll former (Fig. 2-27) is manually operated and consists of three rolls, two housings, a base, and a handle. The handle turns the two front rolls through a system of gears enclosed in the housing. By properly adjusting the roller spacing, metal can be formed into a curve.

Figure 2-26   Sheet-metal brake (box-and-pan brake).

Grinders

A grinder is a cutting tool with a large number of cutting edges arranged so that when they become dull they break off and new cutting edges take their place.

Silicon carbide and aluminum oxide are the abrasives used in most grinding wheels. Silicon carbide is the cutting agent to grind hard, brittle material, such as cast iron. It is also used to grind aluminum, brass, bronze, and copper. Aluminum oxide is

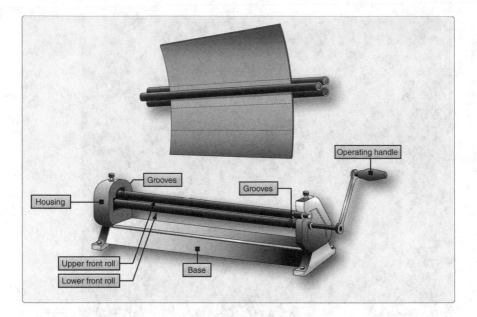

Figure 2-27 Slip roll former.

the cutting agent to grind steel and other metals of high tensile strength.

The size of the abrasive particles used in grinding wheels is indicated by a number that corresponds to the number of meshes per linear inch in the screen, through which the particles will pass. As an example, a #30 abrasive will pass through a screen with 30 holes per linear inch, but will be retained by a smaller screen, with more than 30 holes per linear inch.

A common bench grinder, found in most metalworking shops, is shown in Fig. 2-28. This grinder can be used to dress mushroomed heads on chisels, and points on chisels, screwdrivers, and drills. It can be used to remove excess metal from work and to smooth metal surfaces.

As a rule, it is not good practice to grind work on the side of an abrasive wheel. When an abrasive wheel becomes worn, its cutting efficiency is reduced because of a decrease in surface speed. When a wheel becomes worn in this manner, it should be discarded and a new one installed.

Before using a bench grinder, the abrasive wheels should be checked to be sure that they are firmly held on the spindles by the flange nuts. If an abrasive wheel flies off or becomes loose, it could seriously injure the operator, in addition to ruining the grinder.

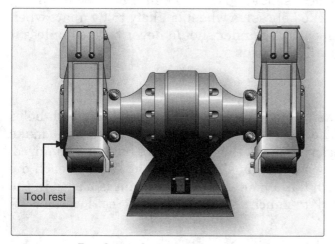

Tool rest

Figure 2-28 Bench grinder.

Figure 2-29 Rotary punch.

Another hazard is loose tool rests. A loose tool rest could cause the tool or piece of work to be "grabbed" by the abrasive wheel and cause the operator's hand to come in contact with the wheel.

Goggles should always be worn when using a grinder—even if eye shields are attached to it. Goggles should fit firmly against your face and nose. This is the only way to protect your eyes from the fine pieces of steel.

The abrasive wheel should be checked for cracks before using the grinder. A cracked abrasive wheel is likely to fly apart when turning at high speeds. A grinder should never be used unless it is equipped with wheel guards.

Rotary punch

Figure 2-29 shows a rotary punch which is used to punch holes in metal parts. The rotary punch can cut radii in corners, make washers, and perform many other jobs where holes are required. The diameter of the punch is stamped on the front of each die holder. Each punch has a point in its center that is placed in the center punch mark to punch the hole in the correct location.

3

Materials and Fabricating

Many different materials go into the manufacture of an aerospace vehicle. Some of these materials are:

- Aluminum and aluminum alloys
- Titanium and titanium alloys
- Magnesium and magnesium alloys
- Steel and steel alloys

Aluminum and Aluminum Alloys

Aluminum is one of the most widely used metals in modern aircraft construction. It is light weight, yet some of its alloys have strengths greater than that of structural steel. It has high resistance to corrosion under the majority of service conditions. The metal can easily be worked into any form and it readily accepts a wide variety of surface finishes.

Being light weight is perhaps aluminum's best-known characteristic. The metal weighs only about 0.1 pound per cubic inch, as compared with 0.28 for iron.

Commercially pure aluminum has a tensile strength of about 13,000 pounds per square inch. Its usefulness as a structural material in this form, thus, is somewhat limited. By working the metal, as by cold rolling, its strength can be approximately doubled. Much larger increases in strength can be obtained by alloying aluminum with small percentages of one or more other metals, such as manganese, silicon, copper, magnesium, or zinc. Like pure aluminum, the alloys are also made stronger by cold working. Some of the alloys are further strengthened and hardened by heat treatments. Today, aluminum alloys with tensile strengths approaching 100,000 pounds per square inch are available.

A wide variety of mechanical characteristics, or tempers, is available in aluminum alloys through various combinations of cold work and heat treatment. In specifying the temper for any given product, the fabricating process and the amount of cold work to which it will subject the metal should be kept in mind. In other words, the temper specified should be such that the amount of cold work that the metal will receive during fabrication will develop the desired characteristics in the finished products.

When aluminum surfaces are exposed to the atmosphere, a thin invisible oxide skin forms immediately that protects the metal from further oxidation. This self-protecting characteristic gives aluminum its high resistance to corrosion. Unless exposed to some substance or condition that destroys this protective oxide coating, the metal remains fully protected against corrosion. Some alloys are less resistant to corrosion than others, particularly certain high-strength alloys. Such alloys in some forms can be effectively protected from the majority of corrosive influences, however, by cladding the exposed surface or surfaces with a thin layer of either pure aluminum or one of the more highly corrosion-resistant alloys. Trade names for some of the clad alloys are *Alclad* and *Pureclad*.

The ease with which aluminum can be fabricated into any form is one of its most important assets. The metal can be cast by any method known to foundry-men; it can be rolled to any desired thickness down to foil thinner than paper; aluminum sheet can be stamped, drawn, spun, or roll-formed. The metal also can be hammered or forged. There is almost no limit to the different shapes in which the metal might be extruded.

The ease and speed that aluminum can be machined is one of the important factors contributing to the use of finished aluminum parts. The metal can be turned, milled, bored, or machined at the maximum speeds of which the majority of machines are capable. Another advantage of its flexible machining characteristics is that aluminum rod and bar can readily be used in the high-speed manufacture of parts by automatic screw machines.

Almost any method of joining is applicable to aluminum, riveting, welding, brazing, or soldering. A wide variety of mechanical aluminum fasteners simplifies the assembly of many products. Adhesive bonding of aluminum parts is widely used in joining aircraft components.

Alloy and temper designations

Aluminum alloys are available in the cast and wrought form. Aluminum castings are produced by pouring molten aluminum alloy into sand or metal molds. Aluminum in the wrought form is obtained three ways:

- Rolling slabs of hot aluminum through rolling mills that produce sheet, plate, and bar stock.

- Extruding hot aluminum through dies to form channels, angles, T sections, etc.

- Forging or hammering a heated billet of aluminum alloy between a male and female die to form the desired part.

Cast and wrought aluminum alloy designation system

A system of four-digit numerical designations is used to identify wrought aluminum and wrought aluminum alloys. The first digit indicates the alloy group, as follows:

Aluminum, 99.00 percent minimum and greater	1xxx
Aluminum alloys grouped by major alloying elements	
Copper	2xxx
Manganese	3xxx
Silicon	4xxx
Magnesium	5xxx
Magnesium and Silicon	6xxx
Zinc	7xxx
Other element	8xxx
Unused series	9xxx

The second digit indicates modifications of the original alloy or impurity limits. The last two digits identify the aluminum alloy

Alloy	Percentage of alloying elements (aluminum and normal impurities constitute remainder)								
	Copper	Silicon	Manganese	Magnesium	Zinc	Nickel	Chromium	Lead	Bismuth
1100	—	—	—	—	—	—	—	—	—
3003	—	—	1.2	—	—	—	—	—	—
2011	5.5	—	—	—	—	—	—	0.5	0.5
2014	4.4	0.8	0.8	0.4	—	—	—	—	—
2017	4.0	—	0.5	0.5	—	—	—	—	—
2117	2.5	—	—	0.3	—	—	—	—	—
2018	4.0	—	—	0.5	—	2.0	—	—	—
2024	4.5	—	0.6	1.5	—	—	—	—	—
2025	4.5	0.8	0.8	—	—	—	—	—	—
4032	0.9	12.5	—	1.0	—	0.9	—	—	—
6151	—	1.0	—	0.6	—	—	0.25	—	—
5052	—	—	—	2.5	—	—	0.25	—	—
6053	—	0.7	—	1.3	—	—	0.25	—	—
6061	0.25	0.6	—	1.0	—	—	0.25	—	—
7075	1.6	—	—	2.5	5.6	—	0.3	—	—

Figure 3-1   Nominal composition of wrought aluminum alloys.

or indicate the aluminum purity. Figure 3-1 shows the percentage of alloying elements in common aluminum alloys.

Aluminum

In the first group (1xxx) for minimum aluminum purities of 99.00 percent and greater, the last two of the four digits in the designation indicate the minimum percentage. Because of its low strength, pure aluminum is seldom used in aircraft.

Aluminum alloys

In the 2xxx through 8xxx alloy groups, the last two of the four digits in the designation have no special significance, but serve only to identify the different aluminum alloys in the group. The second digit in the alloy designation indicates alloy modifications. If the second digit in the designation is zero, it indicates the original alloy; integers 1 through 9, which are assigned consecutively, indicate alloy modifications.

Temper designation system

Where used, the temper designation follows the alloy designation and is separated from it by a dash: 7075-T6, 2024-T4, etc. The temper designation consists of a letter that indicates the basic temper that can be more specifically defined by the addition of one or more digits. Designations are shown in Fig. 3-2.

Characteristics of Aluminum Alloys

In high-purity form, aluminum is soft and ductile. Most aircraft uses, however, require greater strength than pure aluminum

Nonheat-Treatable Alloys		Heat-Treatable Alloys	
Temper designation	Definition	Temper designation	Definition
–O	Annealed recrystallized (wrought products only) applies to softest temper of wrought products.	–T42	Solution heat-treated by the user regardless of prior temper (applicable only to 2014 and 2024 alloys).
–H12	Strain-hardened one-quarter-hard temper.	–T5	Artificially aged only (castings only).
–H14	Strain-hardened half-hard temper.	–T6	Solution heat-treated and artificially aged.
–H16	Strain-hardened three-quarters-hard temper.	–T62	Solution heat-treated and aged by user regardless of prior temper
–H18	Strain-hardened full-hard temper.		(applicable only to 2014 and 2024
–H22	Strain-hardened and partially annealed to one-quarter-hard temper.	–T351, –T451 –T3510, –T3511, –T4510, –T4511.	alloys). Solution heat-treated and stress re- lieved by stretching to produce a permanent set of 1 to 3 percent, depending on the product.
–H24	Strain-hardened and partially annealed to half-hard temper.		
–H26	Strain-hardened and partially annealed to three-quarters-hard temper.	–T651, –T851, –T6510, –T8510, –T6511, –T8511.	Solution heat-treated, stress relieved by stretching to produce a perm- anent set of 1 to 3 percent, and artifically aged.
–H28	Strain-hardened and partially annealed to full-hard temper.	–T652	Solution heat-treated, compressed to
–H32	Strain-hardened and then stabilized. Final temper is one-quarter hard.		produce a permanent set and then artificially aged.
–H34	Strain-hardened and then stabilized. Final temper is one-half hard.	–T81	Solution heat-treated, cold-worked by flattening or straightening operation, and then artificially
–H36	Strain-hardened and then stabilized. Final temper is three-quarters hard.		aged.
–H38	Strain-hardened and then stabilized. Final temper is full-hard.	–T86	Solution heat-treated, cold-worked by reduction of 6 percent, and then artificially aged.
–H112	As fabricated; with specified mechanical property limits.	–F	For wrought alloys; as fabricated. No mechanical properties limits.
–F	For wrought alloys; as fabricated. No mechanical properties limits. For cast alloys; as cast.		For cast alloys; as cast.
–O	Annealed recrystallized (wrought products only) applies to softest temper of wrought products.		
–T2	Annealed (castings only.)		
–T3	Solution heat-treated and cold-worked by the flattening or straightening operation.		
–T36	Solution heat-treated and cold-worked by reduction of 6 percent.		
–T4	Solution heat-treated.		

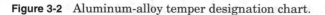

Figure 3-2 Aluminum-alloy temper designation chart.

affords. This is achieved in aluminum first by the addition of other elements to produce various alloys, which singly or in combination impart strength to the metal. Further strengthening is possible by means that classify the alloys roughly into two categories, nonheat treatable and heat treatable.

Nonheat-treatable alloys

The initial strength of alloys in this group depends upon the hardening effect of elements, such as manganese, silicon, iron, and magnesium, singly or in various combinations. The nonheat-treatable alloys are usually designated, therefore, in the 1000, 3000, 4000, or 5000 series. Because these alloys are work-hardenable, further strengthening is made possible by various degrees of cold working, denoted by the "H" series of tempers. Alloys containing appreciable amounts of magnesium when supplied in strain-hardened tempers are usually given a final elevated-temperature treatment, called *stabilizing*, to ensure stability of properties.

Heat-treatable alloys

The initial strength of alloys in this group is enhanced by the addition of such alloying elements as copper, magnesium, zinc, and silicon. Because these elements singly or in various combinations show increasing solid solubility in aluminum with increasing temperature, it is possible to subject them to thermal treatments that will impart pronounced strengthening.

The first step, called *heat treatment* or *solution heat treatment*, is an elevated-temperature process designed to put the soluble element or elements in solid solution. This is followed by rapid quenching, usually in water, which momentarily "freezes" the structure and, for a short time, renders the alloy very workable; selected fabricators retain this more-workable structure by storing the alloys at below-freezing temperatures until initiating the formation process. Ice box rivets are a typical example. At room or elevated temperatures, the alloys are unstable after quenching, however, and precipitation of the constituents from the supersaturated solution begins.

After a period of several days at room temperature, termed *aging* or *room-temperature precipitation*, the alloy is considerably

stronger. Many alloys approach a stable condition at room temperature, but selected alloys, particularly those containing magnesium and silicon or magnesium and zinc continue to age-harden
for long periods of time at room temperature.

By heating for a specified time at slightly elevated temperatures, even further strengthening is possible and properties are
stabilized, called *artificial aging* or *precipitation hardening*. By
the proper combination of solution heat treatment, quenching,
cold working, and artificial aging, the highest strengths are
obtained.

Clad alloys

The heat-treatable alloys in which copper or zinc are major alloying constituents are less resistant to corrosive attack than the
majority of nonheat-treatable alloys. To increase the corrosion
resistance of these alloys in sheet and plate form, they are often
clad with high-purity aluminum, a low magnesium-silicon alloy,
or an alloy that contains 1 percent zinc. The cladding, usually
from 2½ to 5 percent of the total thickness on each side, not only
protects the composite because of its own inherently excellent
corrosion resistance, but also exerts a galvanic effect that further
protects the core material.

Annealing characteristics

All wrought aluminum alloys are available in annealed form. In
addition, it might be desirable to anneal an alloy from any other
initial temper, after working, or between successive stages of
working, such as deep drawing.

Most aluminum alloys are heated to a temperature of 750 to
775°F for two hours in a furnace. After two hours, the furnace
will be shut off and the alloy will be allowed to slowly cool in the
furnance.

Typical uses of aluminum and its alloys

Various aluminum alloys are used for aircraft fabrication:

- *1000 series* Aluminum of 99 percent or higher purity has practically no application in the aerospace industry. These alloys

are characterized by excellent corrosion resistance, high thermal and electrical conductivity, low mechanical properties, and excellent workability. Moderate increases in strength can be obtained by strain hardening. Soft, 1100 rivets are used in nonstructural applications.

- *2000 series* Copper is the principal alloying element in this group. These alloys require solution heat treatment to obtain optimum properties; in the heat-treated condition, mechanical properties are similar to, and sometimes exceed, those of mild steel. In some instances, artificial aging is used to further increase the mechanical properties. This treatment materially increases yield strength. These alloys in the form of sheet are usually clad with a high-purity alloy. Alloy 2024 is perhaps the best known and most widely used aircraft alloy. Most aircraft rivets are of alloy 2117.

- *3000 series* Manganese is the major alloying element of alloys in this group, which are generally nonheat-treatable. One of these is 3003, which has limited use as a general-purpose alloy for moderate-strength applications that require good workability, such as cowlings and nonstructural parts. Alloy 3003 is easy to weld.

- *4000 series* This alloy series is seldom used in the aerospace industry.

- *5000 series* Magnesium is one of the most effective and widely used alloying elements for aluminum. When it is used as the major alloying element, or with manganese, the result is a moderate- to high-strength nonheat-treatable alloy. Alloys in this series possess good welding characteristics and good resistance to corrosion in various atmospheres. It is widely used for the fabrication of tanks and fluid lines.

- *6000 series* Alloys in this group contain silicon and magnesium in approximate proportions to form magnesium silicide, thus making them heat-treatable. The major alloy in this series is 6061, one of the most versatile of the heat-treatable alloys. Although less strong than most of the 2000 or 7000 alloys, the magnesium-silicon (or magnesium-silicide) alloys possess good formability and corrosion resistance, with medium strength.

- *7000 series* Zinc is the major alloying element in this group. When coupled with a smaller percentage of magnesium, the results are heat-treatable alloys with very high strength. Usually other elements, such as copper and chromium, are also added in small quantities. The outstanding member of this group is 7075, which is among the highest-strength alloys available and is used in airframe structures and for highly stressed parts.

Heat treatment of aluminum alloys

The heat treatment of aluminum alloys is summarized in Fig. 3-3.

There are several heat-treating processes to improve the strength or workability of aluminum alloys. These processes are solution heat treatment, artificial aging (also called precipitation heat treatment), and annealing.

The hardening of an aluminum alloy by heat treatment consists of four steps:

1. Heating to a predetermined temperature.

2. Soaking at temperature for a specified length of time.

3. Rapidly quenching to a relatively low temperature.

4. Aging or precipitation hardening at room temperature or as a result of a low-temperature heat treatment.

The first three steps above are known as solution heat treatment. Some alloys like the 2000 series of aluminum alloys will age at room temperature, but the 7000 series of alloys remain unstable and will need an additional heat treatment to age. This is called

Alloy	Solution heat treatment			Precipitation heat treatment		
	Temperature (°F)	Quench	Temperature designation	Temperature (°F)	Time of aging	Temperature designation
2017	930–950	Cold water	T4			T
2117	930–950	Cold water	T4			T
2024	910–930	Cold water	T4			T
6053	960–980	Water	T4	445–455	1–2 hours or	T5
				345–355	8 hours	T6
6061	960–980	Water	T4	315–325	18 hours or	T6
				345–355	8 hours	T6
7075	870	Water		250	24 hours	T6

Figure 3-3 Conditions for heat treatment of aluminum alloys. Heating times vary with the product, type of furnace, and thickness of material. Quenching is normally in cold water, although hot water or air blasting can be used for bulky sections. For information only. Not to be used for actual heat treatment.

artificial aging or precipitation heat treatment. Figure 3-3 shows the solution and precipitation heat-treatment conditions.

Identification of aluminum

To provide a visual means to identify the various grades of aluminum and aluminum alloys, these metals are usually marked with such symbols as Government Specification Number, the temper or condition furnished, or the commercial code marking. Plate and sheet are usually marked with specification numbers or code markings in rows approximately six inches apart. Tubes, bars, rods, and extruded shapes are marked with specification numbers or code markings continuously or at intervals of 3 to 5 feet along the length of each piece. The commercial code marking consists of a number that identifies the particular composition of the alloy. In addition, letter suffixes designate the temper designation. See Fig. 3-4.

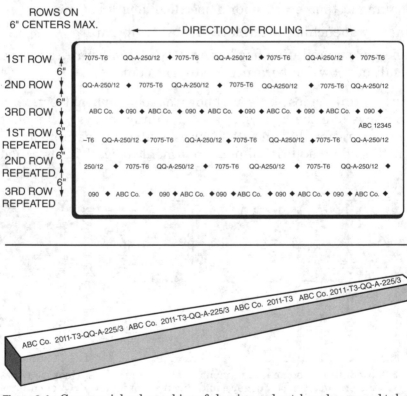

Figure 3-4   Commercial code marking of aluminum sheet, bar, shapes, and tubes.

Handling Aluminum

The surface of "clad" aluminum alloy is very soft and scratches easily. Special care must be used when handling this material. Some suggestions include:

- Keep work area and tables clean.
- Lift material from surface to move it. Do not slide material.
- Keep tools and sharp objects off the surface unless necessary for trimming, drilling, or holding.
- Do not stack sheets of metal together unless interleaved with a neutral kraft paper.
- Prevent moisture from accumulating between sheets.
- Protect material, as necessary, to prevent damage when transporting on "A" frames.

Forming Aluminum Alloys

Forming at the factory

Present-day aircraft manufacturers maintain service departments that include complete spare parts inventories. Detailed parts catalogs are available for all aircraft, including individual wing ribs and pilot-drilled skin panels, for example. For this reason, it is normally not necessary for the field mechanic to be skilled in all phases of sheet-metal forming. It is more cost effective to procure parts from the factory, rather than fabricate them from scratch.

Although the field mechanic might not be required to fabricate individual parts, he should be familiar with the forming processes used by the factory. Also, he will be required to fabricate complete assemblies from factory-supplied parts during repair operations.

Parts are formed at the factory on large presses or by drop hammers equipped with dies of the correct shape. Every part is planned by factory engineers, who set up specifications for the materials to be used so that the finished part will have the correct temper when it leaves the machines. A layout for each part is prepared by factory draftsmen.

The verb *form* means to shape or mold into a different shape or in a particular pattern, and thus would include even casting.

However, in most metal-working terminology, "forming" is generally understood to mean changing the shape by bending and deforming solid metal.

In the case of aluminum, this is usually at room temperature. In metal-working, "forming" includes bending, brake forming, stretch forming, roll forming, drawing, spinning, shear forming, flexible die forming, and high-velocity forming.

Other "forming" methods, such as machining, extruding, forging, and casting do change the shape of the metal, by metal removal or at elevated temperatures. However, these processes use different tooling and/or equipment.

Manufacturers form aluminum by rolling, drawing, extruding, and forging to create the basic aluminum shapes from which the metalworker, in turn, makes all types of end products. As a group, the aluminum products fabricated from ingot by the producers are called *mill products*.

The principal mill products utilized by the metalworker in forming are sheet, plate, rod, bar, wire, and tube. Sheet thickness ranges from 0.006 through 0.249 inch; plate is 0.250 inch or more; rod is ⅜-inch diameter or greater; bar is rectangular, hexagonal, or octagonal in cross section, having at least one perpendicular distance between faces of ⅜ inch or greater. Wire is 0.374 inch or less.

Most parts are formed without annealing the metal, but if extensive forming operations, such as deep draws (large folds) or complex curves, are planned, the metal is in the dead-soft or annealed condition. During the forming of some complex parts, operations might have to be stopped and the metal annealed before the process can be continued or completed. Alloy 2024 in the "O" condition can be formed into almost any shape by the common forming operations, but it must be heat treated afterward.

Blanking

Blanking is a cutting operation that produces a blank of the proper size and shape to form the desired product. Sawing, milling, or routing, are generally used to produce large or heavy-gauge blanks. Routing is the most common method used in the aerospace industry to produce blanks for forming.

Bending

Light-gauge aluminum is easily bent into simple shapes on the versatile hand-operated bending brake. This machine also is commonly known by several other names, including *apron* or *leaf brake*, *cornice brake*, *bar folder*, or *folding brake* (see Chap. 2). More complex shapes are formed by bending on press brakes fitted with proper dies and tooling.

Allowances must be made for springback in bending age-hardened or work-hardened aluminum. Soft alloys of aluminum have comparatively little springback. Where springback is a factor, it is compensated by "overforming" or bending the material beyond the limits actually desired in the final shape. Thick material springs back less than thinner stock in a given alloy and temper.

The proper amount of overforming is generally determined by trial, then controlled by the metalworker in hand or bending brake operations. In press-brake bending, springback is compensated for by die and other tool design, use of adjustable dies, or adjustment of the brake action.

Press-Brake Forming

Hydraulic and mechanical presses are used to form aluminum (and other metals) into complex shapes. Precisely shaped mating dies of hardened tool steel, are made in suitable lengths to produce shapes in one or more steps or passes through the press. The dies are changed as required. See Fig. 3-5.

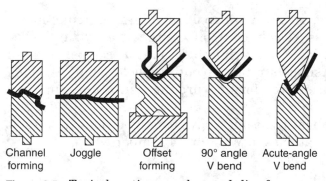

| Channel forming | Joggle | Offset forming | 90° angle V bend | Acute-angle V bend |

Figure 3-5 Typical mating punches and dies for press-brake work; cross section of the formed shape is indicated for each operation. Punch and die are as long as required for workpiece and press capacity.

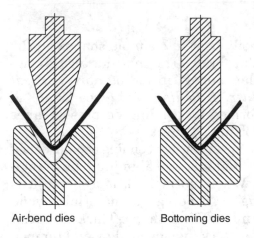

Air-bend dies   Bottoming dies

Figure 3-6 Air-bend and bottoming dies.

Bends made on press brakes usually are done either by the air-bending or by the bottoming method. In air bending, the punch has an acute angle between 30 and 60 degrees, thus providing enough leeway so that for many bends springback compensation can be made by press adjustments alone. See Fig. 3-6. The term *air bending* is derived from the fact that the workpiece spans the gap between the nose of the punch and the edges of the ground die.

In bottoming, the workpiece is in contact with the complete working surfaces of both punch and die, and accurate angular tolerances can thus be obtained. Bottoming requires three to five times greater pressure than air bending.

Stretch Forming

Compound curves, accurate dimensions, minimum reduction in material thickness, closely controlled properties, wrinkle-free shapes, and sometimes cost savings over built-up components can be achieved in a single stretch beyond its yield point. Airplane skins are typical stretch-formed products in aluminum. See Fig. 3-7.

Forming of nonheat-treatable alloys usually is done in the soft O temper; heat-treatables in W, O, or T4 tempers.

Hydro Press Forming

Seamless, cup-like aluminum shapes are formed without wrinkles or drastically altering original metal thickness, on standard

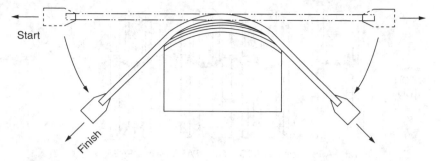

Figure 3-7   One type of stretch forming where the work is stretched over a fixed male die.

single-action presses for most shallow shells, and on double-action presses for deeper and more difficult draws. Both mechanical or hydraulic power is used, the latter offering more control, which is particularly advantageous for deep and some complex shapes. The part is formed between a male and female die attached to hydropress bed or platen and the hydraulic actuated ram, respectively.

Roll Forming

A series of cylindrical dies in sets of two—male and female—called *roll sets* are arranged in the roll-bending machine so that sheet or plate is progressively formed to the final shape in a continuous operation. See Fig. 3-8. By changing roll sets, a wide variety of aluminum products, including angles and channels, such as used for stringers, can be produced at high production rates of 100 feet per minute and faster.

Flexible-Die Forming

Under high pressure, rubber and similar materials act as a hydraulic medium, exerting equal pressure in all directions. In drawing, rubber serves as an effective female die to form an aluminum blank around a punch or form block that has been contoured to the desired pattern. The rubber exerts (transmits) the pressure because it resists deformation; this serves to control local elongation in the aluminum sheet being formed. See Fig. 3-9.

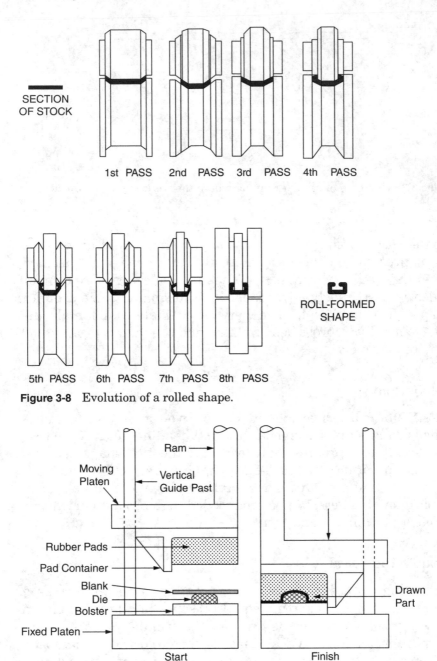

Figure 3-8   Evolution of a rolled shape.

Figure 3-9   Flexible die-forming process forms parts using rubber pad "punch" in large hydraulic presses ranging from 1000 to 10,000 tons capacity. Note: Rubber pad is really the "die" and depth to which any part can be formed is a function of the thickness of the pad and many other variables.

Use of rubber pads for the female die greatly reduces die costs, simplifies machine setup, reduces tool wear and eliminates die marks on the finished product. Identical parts, but in different gauges of material, can be made without making tool changes.

Several flexible-die processes are used to form aluminum. Although the operating details vary, these processes can be classified under two broad categories:

- *Shallow-draw methods* rely on the pressure exerted against the rubber pad to hold the blank as well as form the part.

- *Deeper-draw methods* have independent blank-holding mechanisms.

Machining

Lathes, drills, milling cutters, and other metal-removal machines commonly found in metalworking shops are routinely used to shape aluminum alloys.

For aluminum, cutting speeds are generally much higher than for other metals; the cutting force required is low, the as-cut finish is generally excellent, the dimensional control is good, and the tool life is outstanding.

Single-point tools are used to turn, bore, plane, and shape. In turning and boring, the work generally is rotated while the cutting tool remains stationary; however, when boring is performed on a milling machine or boring mill, the tool rotates and the work is stationary. In planing, the work moves and indexes while the tool is stationary; in shaping, the work is fixed and the tool moves.

Drilling

Drilling is covered in detail in Chap. 4.

Turret Lathes and Screw Machines

Multi-operation machining is carried out in a predetermined sequence on turret lathes, automatic screw machines, and similar equipment. Speeds and feeds are generally near or at upper limits for each type of cutting, with each new operation following in rapid sequence the one just completed.

Automatic screw machines mass produce round solid and hollow parts (threaded and/or contoured) from continuously fed bar or rod, using as many as eight or more successive (and some simultaneous) operations on a variety of complex-tooled turrets, cross-slides, cutting attachments, and stock-feeding devices.

Milling

Aluminum is one of the easiest metals to shape by milling. High spindle speeds and properly designed cutters, machines, fixtures, and power sources can make cuts in rigid aluminum workpieces at high rates of speed.

Milling machines range in size from small, pedestal-mounted types to spar and skin mills with multiple cutting heads and individual motor drives, mounted on gantries that run on the entire 200- to 300-foot length of the machines' beds. These latter machines use computer numeric control (CNC) and are capable of complex contour milling while holding remarkably close tolerances over entire lengths of the part.

Routing

Routers used for machining aluminum have evolved from similar equipment originally and currently used in woodworking. These machines include portable hand routers, hinged and radial routers, and profile routers. Both plain and carbide-tipped high-speed steel tools, rotating at 20,000 rpm (or faster), are used.

The principal router applications for aluminum are for edge-profiling shapes from single or stacked sheet or plate, and for area removal of any volume of metal when the router is used as a skin or spar mill.

Forging

Hammering or squeezing a heated metal into a desired shape is one of the oldest metalworking procedures; such "forging" was one of the first fabricating processes used for making things of aluminum.

Die forgings, also called *close die forgings*, are produced by hammering or squeezing the metal between a suitable punch and

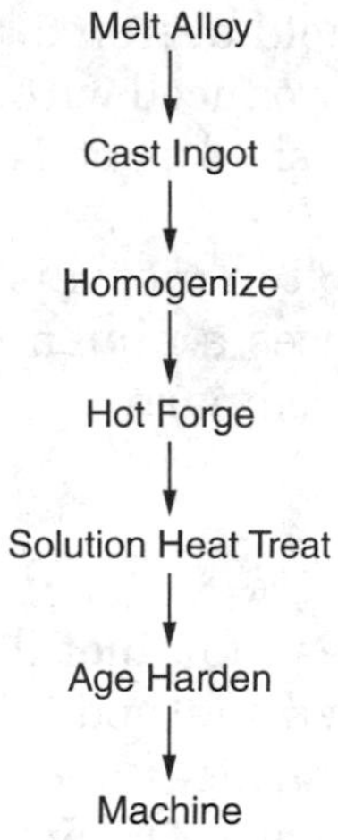

Melt Alloy
↓
Cast Ingot
↓
Homogenize
↓
Hot Forge
↓
Solution Heat Treat
↓
Age Harden
↓
Machine

Figure 3-10 Fabrication sequence for an airplane landing gear.

die set. Excellent accuracy and detail are attained and advantageous grain-flow patterns are established, imparting maximum strength to the alloy used.

Consider, as an example, the manufacture of an airplane landing gear part from alloy 7075. This alloy basically contains 5.5 percent zinc, 2.5 percent magnesium, and 1.5 percent copper, and is age hardenable.

Refer to the flow chart (Fig. 3-10). The alloy is prepared by melting, and an ingot is cast. The ingot is homogenized, and then hot forged between two dies of the desired shape. The finished forging is solution heat treated at about 900°F and quenched in water.

After solution heat treating, it is age hardened at about 250°F. Some final surface machining completes the part and it is ready to assemble on the airplane.

Casting

Three basic casting processes are: sand casting, permanent mold casting, and die casting.

Sand casting uses a mold made from sand, based on the use of a pattern. The mold is destroyed when the cast part is removed. Sand castings are used for small-quantity runs. The finished casting has a rough surface and usually requires some machining.

Permanent mold casting utilizes a permanent mold of iron or steel that can be used repeatedly. A finished part is produced with smooth surfaces. Dimensional accuracy of the finished part is close to that of a die-cast part.

Die casting uses a permanent mold, whereby molten metal is forced into the die cavity under pressure. It produces a dimensionally accurate, thin-sectioned and smooth-surfaced part.

Chemical Milling

Chemical milling is a dimensional etching process for metal removal. In working aluminum, it is the preferred method of removing less than 0.125 inch from large, intricate surfaces, such as integrally stiffened wing skins for high-performance aircraft. Sodium-hydroxide-base or other suitable alkaline solutions are generally used to chemically mill aluminum. Process is carried out at elevated temperatures. Metal removal (dissolution) is controlled by masking, rate of immersion, duration of immersion, and the composition and temperature of bath.

Dissolution of a 0.01-inch thickness of aluminum per minute is a typical removal rate. Economics dictates the removal of thicknesses greater than 0.250 inch by mechanical means. The choice of method between the aforementioned 0.125- and 0.250-inch metal-removal thickness depends on the fillet ratio and weight penalty.

Making Straight-Line Bends

Forming at the factory (as covered in the previous section) involves specialized equipment and techniques. Therefore, it is generally beyond the scope of the field mechanic. However, an example of straight-line bends is appropriate.

When forming straight bends, the thickness of the material, its alloy composition, and its temper condition must be considered. Generally speaking, the thinner the material, the sharper it can be bent (the smaller the radius of bend), and the softer the material, the sharper the bend. Other factors that must be considered when making straight-line bends are bend allowance, setback, and the brake or sight line.

The radius of bend of a sheet of material is the radius of the bend, as measured on the inside of the curved materials. The minimum radius of bend of a sheet of material is the sharpest curve, or bend, to which the sheet can be bent without critically weakening the metal at the bend. If the radius of bend is too small, stresses and strains will weaken the metal and could result in cracking.

A minimum radius of bend is specified for each type of aircraft sheet metal. The kind of material, thickness, and temper condition of the sheet are factors that affect the minimum radius. Annealed sheet can be bent to a radius approximately equal to its thickness. Stainless steel and 2024-T aluminum alloy require a fairly large bend radius.

A general rule for minimum bend radii is:

- $1 \times thickness$ for O temper.
- $2\frac{1}{2} \times thickness$ for T4 temper.
- $3 \times thickness$ for T3 temper.

Bend allowance

When making a bend or fold in a sheet of metal, the bend allowance must be calculated. Bend allowance is the length of material required for the bend. This amount of metal must be added to the overall length of the layout pattern to ensure adequate metal for the bend (Fig. 3-11).

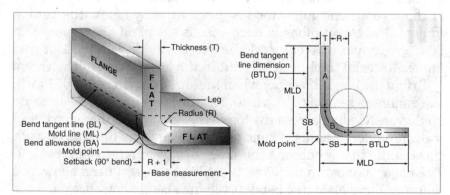

Figure 3-11   Bend-allowance terminology.

Bending a strip compresses the material on the inside of the curve and stretches the material on the outside of the curve. However, at some distance between these two extremes lies a space that is not affected by either force. This is known as the *neutral line* or *neutral axis*. It occurs at a distance approximately 0.445 times the metal thickness $(0.455 \times T)$ from the inside of the radius of the bend.

When bending metal to exact dimensions, the length of the neutral line must be determined so that sufficient material can be allowed for the bend. To save time in calculating the bend allowance, formulas and charts for various angles, radii of bends, material thicknesses, and other factors have been established.

By experimenting with actual bends in metals, aircraft engineers have found that accurate bending results could be obtained by using the following formula for any degree of bend from 1 to 180 degrees:

$$Bend\ allowance = (0.01743 \times R + 0.0078 \times T) \times N$$

where: R = The desired bend radius,
$\quad\quad T$ = Thickness of the material, and
$\quad\quad N$ = Number of degrees of bend.

This formula can be used in the absence of a bend-allowance chart. To determine the bend allowance for any degree of bend by use of the chart (Fig. 3-12), find the allowance per degree for the number of degrees in the bend.

The radius of bend is given as a decimal fraction on the top line of the chart. Bend allowance is given directly below the radius figures. The top number in each case is the bend allowance for a 90-degree angle, whereas the lower-placed number is for a 1-degree angle. Material thickness is given in the left column of the chart.

To find the bend allowance when the sheet thickness is 0.051 inch, the radius of bend is ¼ inch (0.250 inch) and the bend is to be 90 degree. Reading across the top of the bend-allowance chart, find the column for a radius of bend of 0.250 inch. Now, find the block in this column that is opposite the gauge of 0.051 in the column at left. The upper number in the block is 0.428, the correct bend allowance in inches for a 90-degree bend (0.428 inch bend allowance).

If the bend is to be other than 90-degree, use the lower number in the block (the bend allowance for 1-degree) and compute the

Metal Thickness	1/32 .031	1/16 .063	3/32 .094	1/8 .125	5/32 .156	3/16 .188	7/32 .219	1/4 .250	9/32 .281	5/16 .313	11/32 .344	3/8 .375	7/16 .438	1/2 .500
						RADIUS OF BEND, IN INCHES								
.020	.062 .000693	.113 .001251	.161 .001792	.210 .002333	.259 .002874	.309 .003433	.358 .003974	.406 .004515	.455 .005056	.505 .005614	.554 .006155	.603 .006695	.702 .007795	.799 .008877
.025	.066 .000736	.116 .001294	.165 .001835	.214 .002376	.263 .002917	.313 .003476	.362 .004017	.410 .004558	.459 .005098	.509 .005657	.558 .006198	.607 .006739	.705 .007838	.803 .008920
.028	.068 .000759	.119 .001318	.167 .001859	.216 .002400	.265 .002941	.315 .003499	.364 .004040	.412 .004581	.461 .005122	.511 .005680	.560 .006221	.609 .006762	.708 .007862	.805 .007862
.032	.071 .000787	.121 .001345	.170 .001886	.218 .002427	.267 .002968	.317 .003526	.366 .004067	.415 .004608	.463 .005149	.514 .005708	.562 .006249	.611 .006789	.710 .007889	.807 .008971
.038	.075 .00837	.126 .001396	.174 .001937	.223 .002478	.272 .003019	.322 .003577	.371 .004118	.419 .004659	.468 .005200	.518 .005758	.567 .006299	.616 .006840	.715 .007940	.812 .009021
.040	.077 .000853	.127 .001411	.176 .001952	.224 .002493	.273 .003034	.323 .003593	.372 .004134	.421 .004675	.469 .005215	.520 .005774	.568 .006315	.617 .006856	.716 .007955	.813 .009037
.051		.134 .001413	.183 .002034	.232 .002575	.280 .003116	.331 .003675	.379 .004215	.428 .004756	.477 .005297	.527 .005855	.576 .006397	.624 .006934	.723 .008037	.821 .009119
.064		.144 .001595	.192 .002136	.241 .002676	.290 .003218	.340 .003776	.389 .004317	.437 .004858	.486 .005399	.536 .005957	.585 .006498	.634 .007039	.732 .008138	.830 .009220
.072			.198 .002202	.247 .002743	.296 .003284	.436 .003842	.394 .004283	.443 .004924	.492 .005465	.542 .006023	.591 .006564	.639 .007105	.738 .008205	.836 .009287
.078			.202 .002249	.251 .002790	.300 .003331	.350 .003889	.399 .004430	.447 .004963	.496 .005512	.546 .006070	.595 .006611	.644 .007152	.745 .008252	.840 .009333
.081			.204 .002272	.253 .002813	.302 .003354	.352 .003912	.401 .004453	.449 .004969	.498 .005535	.548 .006094	.598 .006635	.646 .007176	.745 .008275	.842 .009357
.091			.212 .002350	.260 .002891	.309 .003432	.359 .003990	.408 .004531	.456 .005072	.505 .005613	.555 .006172	.604 .006713	.653 .007254	.752 .008353	.849 .009435
.094			.214 .002374	.262 .002914	.311 .003455	.361 .004014	.410 .004555	.459 .005096	.507 .005637	.558 .006195	.606 .006736	.655 .007277	.754 .008376	.851 .009458
.102				.268 .002977	.317 .003518	.367 .004076	.416 .004617	.464 .005158	.513 .005699	.563 .006257	.612 .006798	.661 .007339	.760 .008439	.857 .009521
.109				.273 .003031	.321 .003572	.372 .004131	.420 .004672	.469 .005213	.518 .005754	.568 .006312	.617 .006853	.665 .008394	.764 .008493	.862 .009575
.125				.284 .003156	.333 .003697	.383 .004256	.432 .004797	.480 .005338	.529 .005678	.579 .006437	.628 .006978	.677 .007519	.776 .008618	.873 .009700
.156					.355 .003030	.405 .004407	.453 .005038	.502 .005570	.551 .006120	.601 .006679	.650 .007220	.698 .007761	.797 .008860	.895 .009942
.188						.417 .004747	.476 .005288	.525 .005829	.573 .006370	.624 .006928	.672 .007469	.721 .008010	.820 .009109	.917 .010191
.250								.568 .006313	.617 .006853	.667 .007412	.716 .007953	.764 .008494	.863 .009593	.961 .010675

Figure 3-12   Bend-allowance chart.

bend allowance. The lower number in this case is 0.004756. Therefore, if the bend is to be 120-degree, the total bend allowance in inches will be 120×0.004756, which equals 0.5707 inch.

When bending a piece of sheet stock, it is necessary to know the starting and ending points of the bend so that the length of the "flat" of the stock can be determined. Two factors are important in determining this: the radius of bend and the thickness of the material.

Notice that setback is the distance from the bend tangent line to the mold point. The mold point is the point of intersection of the lines that extend from the outside surfaces, whereas the bend tangent lines are the starting and end points of the bend.

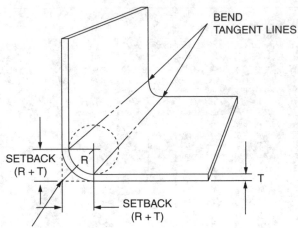

Figure 3-13 Setback, 90° bend.

Also notice that the setback is the same for the vertical flat and the horizontal flat.

To calculate the setback for a 90-degree bend, merely add the inside radius of the bend to the thickness of the sheet stock:

$$Setback = R + T \text{ (Fig. 3-13)}$$

To calculate setback for angles larger or smaller than 90-degree, consult standard setback charts or the K chart (Fig. 3-14) for a value called K, and then substitute this value in the formula:

$$Setback = K (R + T).$$

The value for K varies with the number of degrees in the bend. For example:

Calculate the setback for a 120-degree bend with a radius of bend of 0.125 inch for a sheet 0.032 inch thick;

$$Setback = K(R + T)$$
$$= 1.7320 \, (0.125 + 0.032)$$
$$= 0.272 \text{ inch}$$

Brake or sight line

The brake or sight line is the mark on a flat sheet that is set even with the nose of the radius bar of the cornice brake and serves as

Degree	K	Degree	K	Degree	K	Degree	K	Degree	K
1	0.0087	37	0.3346	73	0.7399	109	1.401	145	3.171
2	0.0174	38	0.3443	74	0.7535	110	1.428	146	3.270
3	0.0261	39	0.3541	75	0.7673	111	1.455	147	3.375
4	0.0349	40	0.3639	76	0.7812	112	1.482	148	3.487
5	0.0436	41	0.3738	77	0.7954	113	1.510	149	3.605
6	0.0524	42	0.3838	78	0.8097	114	1.539	150	3.732
7	0.0611	43	0.3939	79	0.8243	115	1.569	151	3.866
8	0.0699	44	0.4040	80	0.8391	116	1.600	152	4.010
9	0.0787	45	0.4142	81	0.8540	117	1.631	153	4.165
10	0.0874	46	0.4244	82	0.8692	118	1.664	154	4.331
11	0.0963	47	0.4348	83	0.8847	119	1.697	155	4.510
12	0.1051	48	0.4452	84	0.9004	120	1.732	156	4.704
13	0.1139	49	0.4557	85	0.9163	121	1.767	157	4.915
14	0.1228	50	0.4663	86	0.9324	122	1.804	158	5.144
15	0.1316	51	0.4769	87	0.9489	123	1.841	159	5.399
16	0.1405	52	0.4877	88	0.9656	124	1.880	160	5.671
17	0.1494	53	0.4985	89	0.9827	125	1.921	161	5.975
18	0.1583	54	0.5095	90	1.000	126	1.962	162	6.313
19	0.1673	55	0.5205	91	1.017	127	2.005	163	6.691
20	0.1763	56	0.5317	92	1.035	128	2.050	164	7.115
21	0.1853	57	0.5429	93	1.053	129	2.096	165	7.595
22	0.1943	58	0.5543	94	1.072	130	2.144	166	8.144
23	0.2034	59	0.5657	95	1.091	131	2.194	167	8.776
24	0.2125	60	0.5773	96	1.110	132	2.246	168	9.514
25	0.2216	61	0.5890	97	1.130	133	2.299	169	10.38
26	0.2308	62	0.6008	98	1.150	134	2.355	170	11.43
27	0.2400	63	0.6128	99	1.170	135	2.414	171	12.70
28	0.2493	64	0.6248	100	1.191	136	2.475	172	14.30
29	0.2586	65	0.6370	101	1.213	137	2.538	173	16.35
30	0.2679	66	0.6494	102	1.234	138	2.605	174	19.08
31	0.2773	67	0.6618	103	1.257	139	2.674	175	22.90
32	0.2867	68	0.6745	104	1.279	140	2.747	176	26.63
33	0.2962	69	0.6872	105	1.303	141	2.823	177	38.18
34	0.3057	70	0.7002	106	1.327	142	2.904	178	57.29
35	0.3153	71	0.7132	107	1.351	143	2.988	179	114.59
36	0.3249	72	0.7265	108	1.376	144	3.077	180	Inf.

Figure 3-14 K chart.

a guide when bending. The brake line can be located by measuring out one radius from the bend tangent line closest to the end that is to be inserted under the nose of the brake or against the radius form block. The nose of the brake or radius bar should fall directly over the brake or sight line, as shown in Fig. 3-15.

J chart for calculating bend allowance

The J chart can be used to determine the setback and total developed width (TDW) of a flat pattern layout when the inside bend radius, bend angle, and thickness are known. The instructions on how to use the J chart are located on the bottom of the J chart shown in Fig. 3-16.

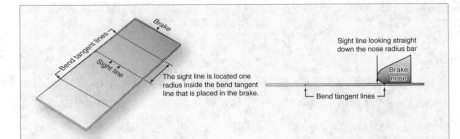

Figure 3-15 Sight line.

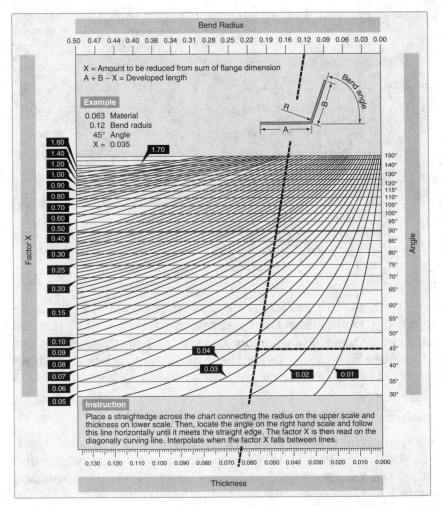

Figure 3-16 J chart.

Making Layouts

It is wise to make a layout or pattern of the part before forming it. This technique reduces wasted material and allows a greater degree of accuracy in the finished part. Where straight-angle bends are concerned, correct allowances must be made for setback and bend allowance.

Relief holes

Wherever two bends intersect, material must be removed to make room for the material contained in the flanges. Holes are therefore drilled at the intersection. These holes, called *relief holes* (Fig. 3-17), prevent strains from being set up at the intersection of the inside-bend tangent lines that would cause the metal to crack. Relief holes also provide a neatly trimmed corner from which excess material can be trimmed.

The size of a relief hole varies with the thickness of the material. The size should be not less than ⅛-inch diameter for aluminum alloy sheet stock up to and including 0.064 inch thick; it should be ³⁄₁₆ inch for stock ranging from 0.072- to 0.128-inch thickness. The most common method of determining the diameter of a relief hole is to use the radius of bend for this dimension, provided that it is not less than the minimum allowance (⅛ inch).

Relief holes must touch the intersection of the inside-bend tangent lines. To allow for possible error in bending, make the relief

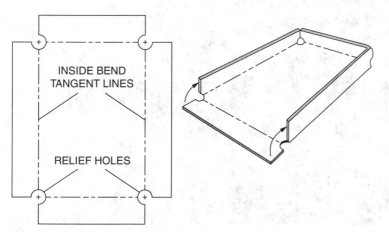

Figure 3-17 Locating relief holes.

holes so that they extend 1⁄32 to 1⁄16 inch behind the inside end tangent lines. The intersection of these lines should be used as the center for the holes. The line on the inside of the curve is cut at an angle toward the relief holes to allow for the stretching of the inside flange.

Miscellaneous shop equipment and procedures

Selected pieces of shop equipment are presented in Chap. 2. Figure 3-18 shows a hand-operated brake for bending sheet metal. Larger brakes are power operated.

Figure 3-18   Hand-operated brake.

Bends of a more complicated design, like a sheet-metal rib having flanges around its contour, should be made over a form block shaped to fit the inside contour of the finished part. Bending the flanges over this die can be accomplished by hand forming, a slow (but practical) method for experimental work (Fig. 3-19).

Machining involves all forms of cutting, whether performed on sheet stock, castings, or extrusions, and involves such operations as shearing (Fig. 3-20), sawing, routing, and lathe and millwork, and such hand operations as drilling, tapping, and reaming.

Magnesium and Magnesium Alloys

Magnesium, the world's lightest structural metal, is a silvery-white material that weighs only two-thirds as much as aluminum. Magnesium does not possess sufficient strength in its pure state for structural uses, but when alloyed with zinc, aluminum, and manganese, it produces an alloy having the highest strength-to-weight ratio of any of the commonly used metals.

Some of today's aircraft require in excess of one-half ton of this metal for use in hundreds of vital spots. Selected wing panels are fabricated entirely from magnesium alloys. These panels weigh 18 percent less than standard aluminum panels and have flown hundreds of satisfactory hours. Among the aircraft parts that have been made from magnesium with a substantial savings in weight are nosewheel doors, flap cover skins, aileron cover skins, oil tanks, floorings, fuselage parts, wingtips, engine nacelles, instrument panels, radio antenna masts, hydraulic fluid tanks, oxygen bottle cases, ducts, and seats.

Magnesium alloys possess good casting characteristics. Their properties compare favorably with those of cast aluminum. In forging, hydraulic presses are ordinarily used, although, under certain conditions, forging can be accomplished in mechanical presses or with drop hammers.

Magnesium alloys are subject to such treatments as annealing, quenching, solution heat treatment, aging, and stabilizing. Sheet and plate magnesium are annealed at the rolling mill. The solution heat treatment is used to put as much of the alloying ingredients as possible into solid solution, which results in high tensile strength and maximum ductility. Aging is applied to castings following heat treatment if maximum hardness and yield strength are desired.

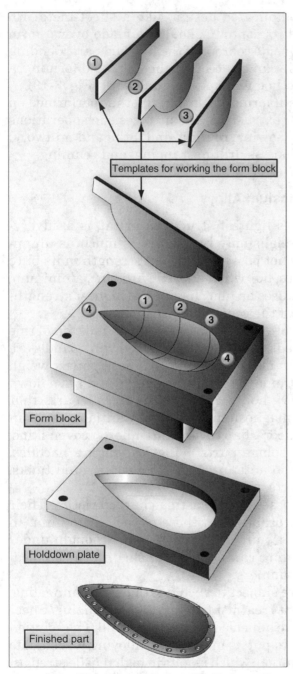

Figure 3-19 Simple form block and hold-down plate for hand forming. A speedier and better production is the use of the hydropress.

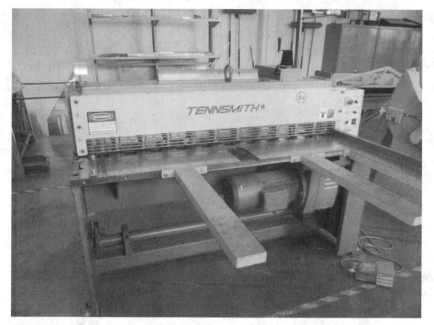

Figure 3-20 Power-operated shear.

Magnesium embodies fire hazards of an unpredictable nature. When in large sections, high thermal conductivity makes it difficult to ignite and prevents it from burning; it will not burn until the melting point is reached, which is 1204°F. However, magnesium dust and fine chips can be ignited easily. Precautions must be taken to avoid this, if possible. If a fire occurs, it can be extinguished with an extinguishing powder, such as powdered soapstone or graphite powder. Water or any standard liquid or foam fire extinguishers cause magnesium to burn more rapidly and can cause explosions.

Magnesium alloys produced in the United States consist of magnesium alloyed with varying proportions of aluminum, manganese, and zinc. These alloys are designated by a letter of the alphabet, with the number 1 indicating high purity and maximum corrosion resistance.

Heat treatment of magnesium alloys

Magnesium alloy castings respond readily to heat treatment, and about 95 percent of the magnesium used in aircraft construction is in the cast form. Heat treatment of magnesium alloy castings is similar to the heat treatment of aluminum alloys because the two types of heat treatment are solution and precipitation (aging). Magnesium, however, develops a negligible change in its properties when allowed to age naturally at room temperatures.

Titanium and Titanium Alloys

In aircraft construction and repair, titanium is used for fuselage skins, engine shrouds, firewalls, longerons, frames, fittings, air ducts, and fasteners. Titanium is used to make compressor disks, spacer rings, compressor blades and vanes, through bolts, turbine housings and liners, and miscellaneous hardware for turbine engines.

Titanium falls between aluminum and stainless steel in terms of elasticity, density, and elevated temperature strength. It has a melting point from 2730 to 3155°F, low thermal conductivity, and a low coefficient of expansion. It is light, strong, and resistant to stress-corrosion cracking. Titanium is approximately 60 percent heavier than aluminum and about 50 percent lighter than stainless steel.

Because of the high melting point of titanium, high-temperature properties are disappointing. The ultimate yield strength of titanium drops rapidly above 800°F. The absorption of oxygen and nitrogen from the air at temperatures above 1000°F makes the metal so brittle on long exposure that it soon becomes worthless. However, titanium does have some merit for short-time exposure up to 3000°F, where strength is not important. Aircraft firewalls demand this requirement.

Titanium is nonmagnetic and has an electrical resistance comparable to that of stainless steel. Some of the base alloys of titanium are quite hard. Heat treating and alloying do not develop the hardness of titanium to the high levels of some of the heat-treated alloys of steel. A heat-treatable titanium alloy was only recently developed. Prior to the development of this alloy, heating and rolling was the only method of forming that could be accomplished. However, it is possible to form the new alloy in the soft condition and heat treat it for hardness.

Iron, molybdenum, and chromium are used to stabilize titanium and produce alloys that will quench harden and age harden.

The addition of these metals also adds ductility. The fatigue resistance of titanium is greater than that of aluminum or steel.

Titanium designations

The A-B-C classification of titanium alloys was established to provide a convenient and simple means to describe titanium alloys. Titanium and titanium alloys possess three basic crystals: A (alpha), B (beta), and C (combined alpha and beta), that have specific characteristics:

- *A (alpha)* All-around performance, good weldability, tough and strong both cold and hot, and resistant to oxidation.

- *B (beta)* Bendability, excellent bend ductility, strong both cold and hot, but vulnerable to contamination.

- *C (combined alpha and beta for compromise performances)* Strong when cold and warm, but weak when hot; good bendability; moderate contamination resistance; and excellent forgeability.

Titanium is manufactured for commercial use in two basic compositions: commercially pure and alloyed. A-55 is an example of a commercially pure titanium; it has a yield strength of 55,000 to 80,000 psi and is a general-purpose grade for moderate to severe forming. It is sometimes used for nonstructural aircraft parts and for all types of corrosion-resistant applications, such as tubing.

Type A-70 titanium is closely related to type A-55, but has a yield strength of 70,000 to 95,000 psi. It is used where higher strength is required, and it is specified for many moderately stressed aircraft parts. For many corrosion applications, it is used interchangeably with type A-55. Type A-55 and type A-70 are weldable.

One of the widely used titanium-base alloys is C-110M. It is used for primary structural members and aircraft skin, has 110,000 psi minimum yield strength, and contains 8 percent manganese.

Type A-110AT is a titanium alloy that contains 5 percent aluminum and 2.5 percent tin. It also has a high minimum yield strength at elevated temperatures with the excellent welding characteristics inherent in alpha-type titanium alloys.

Corrosion characteristics

The corrosion resistance of titanium deserves special mention. The resistance of the metal to corrosion is caused by the formation of a protective surface film of stable oxide or chemi-absorbed oxygen. Film is often produced by the presence of oxygen and oxidizing agents.

Titanium corrosion is uniform. There is little evidence of pitting or other serious forms of localized attack. Normally, it is not subject to stress corrosion, corrosion fatigue, intergranular corrosion, or galvanic corrosion. Its corrosion resistance is equal or superior to 18-8 stainless steel.

Treatment of titanium

Titanium is heat treated for the following purposes:

- Relief of stresses set up during cold forming or machining.
- Annealing after hot working or cold working, or to provide maximum ductility for subsequent cold working.
- Thermal hardening to improve strength.

Working with Titanium

Unlike familiar metals, such as aluminum and steel, which generally require no special techniques and procedures for machining, drilling, tapping or forming, working with titanium requires consideration of its special characteristics. Therefore, a more-detailed discussion of titanium's workability is in order.

Machining of titanium

Titanium can be economically machined on a routine production basis if shop procedures are set up to allow for the physical characteristics common to the metal. The factors that must be given consideration are not complex, but they are vital to the successful handling of titanium.

Most important is that different grades of titanium (i.e., commercially pure and various alloys) will not all have identical machining characteristics. Like stainless steel, the low thermal conductivity of titanium inhibits dissipation of heat within the workpiece itself, thus requiring proper application of coolants.

Generally, good tool life and work quality can be ensured by rigid machine set-ups, use of a good coolant, sharp and proper tools, slower speeds, and heavier feeds. The use of sharp tools is vital because dull tools will accentuate heat build-up to cause undue galling and seizing, leading to premature tool failure.

Milling

The milling of titanium is a more-difficult operation than that of turning. The cutter mills only part of each revolution, and chips tend to adhere to the teeth during that portion of the revolution that each tooth does not cut. On the next contact, when the chip is knocked off, the tooth could be damaged.

This problem can be alleviated to a great extent by using climb milling, instead of conventional milling. In this type of milling, the cutter is in contact with the thinnest portion of the chip as it leaves the cut, minimizing chip "welding."

For slab milling, the work should move in the same direction as the cutting teeth. For face milling, the teeth should emerge from the cut in the same direction as the work is fed.

In milling titanium, when the cutting edge fails, it is usually because of chipping. Thus, the results with carbide tools are often less satisfactory than with cast-alloy tools. The increase in cutting speeds of 20 to 30 percent, which is possible with carbide, does not always compensate for the additional tool-grinding costs. Consequently, it is advisable to try both cast-alloy and carbide tools to determine the better of the two for each milling job. The use of a water-base coolant is recommended.

Turning

Commercially pure and alloyed titanium can be turned with little difficulty. Carbide tools are the most satisfactory for turning titanium. The "straight" tungsten carbide grades of standard designations C1 through C4, such as Metal Carbides C-91 and similar types, provide the best results. Cobalt-type high-speed steels appear to be the best of the many types of high-speed steel available. Cast-alloy tools, such as Stellite, Tantung, Rexalloy, etc., can be used when carbide is not available and when the cheaper high-speed steels are not satisfactory.

Drilling

Successful drilling can be accomplished with ordinary high-speed steel drills. One of the most important factors in drilling titanium is the length of the unsupported section of the drill.

This portion of the drill should be no longer than necessary to drill the required depth of hole and still allow the chips to flow unhampered through the flutes and out of the hole. This permits the application of maximum cutting pressure, as well as rapid removal and reengagement to clear chips, without drill breakage. Use of "Spiro-Point" drill grinding is desirable.

Tapping

The best results in tapping titanium have been with a 65 percent thread. Chip removal is a problem that makes tapping one of the more-difficult machining operations. However, in tapping through-holes, this problem can be simplified by using a gun-type tap with which chips are pushed ahead of the tap. Another problem is the smear of titanium on the land of the tap, which can result in the tap freezing or binding in the hole. An activated cutting oil, such as a sulfurized-and-chlorinated oil, is helpful in avoiding this.

Grinding

The proper combination of grinding fluid, abrasive wheel, and wheel speeds can expedite this form of shaping titanium. Both alundum and silicon carbide wheels are used. The procedure recommended is to use considerably lower wheel speeds than in conventional grinding of steels. A water-sodium nitrite mixture produces excellent results as a coolant. However, this solution can be very corrosive to equipment, unless proper precautions are used.

Sawing

Slow speeds (in the 50-fpm range) and heavy, constant blade pressure should be used. Standard blades should be reground to provide improved cutting efficiency and blade life.

Cleaning after machining

It is recommended that machined parts that will be exposed to elevated temperatures should be thoroughly cleaned to remove all traces of cutting oils. An acceptable recommended solvent is methyl-ethyl-ketone (MEK).

It is advisable not to use low-flash-point cutting oils because the high heat generated during machining could cause the oil to ignite. Water-soluble oils or cutting fluids with a high flash point are recommended.

Shop-forming titanium

Titanium sheet material can be cold or hot formed, although the latter is usually preferable. Forming is best accomplished by one of four basic methods (hydropress, power brake, stretch, or drop hammer), using somewhat more gradual application of pressure than with steel. Titanium mill products are generally shipped in the annealed condition, and thus are in their most workable condition for forming, as received.

Initial forming operations—the preparation of blanks—are much like those used for 18-8 stainless steel: shearing, die blanking, nibbling, and sawing are all satisfactory. To prevent cracks or tears during forming operations of titanium, blanks should be deburred to a round, smooth edge.

Stress relief

As an aid to cold forming, it is usually necessary to stress relieve where more than one stage of fabrication is involved. For example, a part should be stress relieved after brake forming prior to stretching and also between room-temperature hydropress forming stages. After cold-forming operations are complete, heat treatment is necessary to relieve residual stresses imposed during forming.

Ferrous Aircraft Metals

Ferrous applies to the group of metals having iron as their principal constituent.

Identification

If carbon is added to iron, in percentages ranging up to approximately 1 percent, the product is vastly superior to iron alone and is classified as *carbon steel*. Carbon steel forms the base of those alloy steels produced by combining carbon steel with other elements known to improve the properties of steel. A base metal, such as iron, to which small quantities of other metals have been added is called an *alloy*. The addition of other metals changes or improves the chemical or physical properties of the base metal for a particular use.

The steel classification of the SAE (Society of Automotive Engineers) is used in specifications for all high-grade steels used in automotive and aircraft construction. A numerical index system identifies the composition of SAE steels.

Each SAE number consists of a group of digits: the first digit represents the type of steel; the second, the percentage of the principal alloying element; and, usually, the last two or three digits, the percentage, in hundredths of 1 percent, of carbon in the alloy. For example, the SAE number 4130 indicates a molybdenum steel containing 1 percent molybdenum and 0.30 percent carbon.

Type of Steel	Classification
Carbon	1xxx
Nickel	2xxx
Nickel-chromium	3xxx
Molybdenum	4xxx
Chromium	5xxx
Chromium-vanadium	6xxx
Tungsten	7xxx
Silicon-manganese	9xxx

SAE numerical index

Metal stock is manufactured in several forms and shapes, including sheets, bars, rods, tubings, extrusions, forgings, and castings. Sheet metal is made in a number of sizes and thicknesses. Specifications designate thicknesses in thousandths of an inch. Bars and rods are supplied in a variety of shapes, such as round, square, rectangular, hexagonal, and octagonal. Tubing can be obtained in round, oval, rectangular, or streamlined shapes. The size of tubing is generally specified by outside diameter and wall thickness.

The sheet metal is usually formed cold in such machines as presses, bending brakes, drawbenches, or rolls. Forgings are shaped or formed by pressing or hammering heated metal in dies. Castings are produced by pouring molten metal into molds. The casting is finished by machining.

Types, characteristics, and uses of alloyed steels

Steel that contains carbon in percentages range from 0.10 to 0.30 percent is considered *low-carbon steel*. The equivalent SAE numbers range from 1010 to 1030. Steels of this grade are used to make such items as safety wire, selected nuts, cable bushings, or threaded rod ends. This steel, in sheet form, is used for secondary structural parts and clamps, and in tubular form for moderately stressed structural parts.

Steel that contains carbon in percentages that range from 0.30 to 0.50 percent is considered *medium-carbon steel*. This steel is especially adaptable for machining or forging, and where surface hardness is desirable. Selected rod ends and light forgings are made from SAE 1035 steel.

Steel that contains carbon in percentages ranging from 0.50 to 1.05 percent is *high-carbon steel*. The addition of other elements in varying quantities add to the hardness of this steel. In the fully heat-treated condition, it is very hard, will withstand high shear and wear, and will have minor deformation. It has limited use in aircraft. SAE 1095 in sheet form is used to make flat springs and in wire form to make coil springs.

The various nickel steels are produced by combining nickel with carbon steel. Steels containing from 3 to 3.75 percent nickel are commonly used. Nickel increases the hardness, tensile strength, and elastic limit of steel without appreciably decreasing the ductility. It also intensifies the hardening effect of heat treatment. SAE 2330 steel is used extensively for aircraft parts, such as bolts, terminals, keys, clevises, and pins.

Chromium steel has high hardness, strength, and corrosion-resistant properties, and is particularly adaptable for heat-treated forgings that require greater toughness and strength than can be obtained in plain carbon steel. Chromium steel can be used for such articles as the balls and rollers of antifriction bearings.

Chrome-nickel (stainless) steels are the corrosion-resistant metals. The anticorrosive degree of this steel is determined by the surface condition of the metal, as well as by the composition, temperature, and concentration of the corrosive agent.

The principal alloy of stainless steel is chromium. The corrosion-resistant steel most often used in aircraft construction is known as 18-8 steel because it is 18 percent chromium and 8 percent nickel. One distinctive feature of 18-8 steel is that its strength can be increased by coldworking.

Stainless steel can be rolled, drawn, bent, or formed to any shape. Because these steels expand about 50 percent more than mild steel and conduct heat only about 40 percent as rapidly, they are more difficult to weld. Stainless steel can be used for almost any part of an aircraft. Some of its common applications are in the fabrication of exhaust collectors, stacks and manifolds, structural and machine parts, springs, castings, tie rods, and control cables.

Chrome-vanadium steels are made of approximately 18 percent vanadium and about 1 percent chromium. When heat treated, they have strength, toughness, and resistance to wear and fatigue. A special grade of this steel in sheet form can be cold formed into intricate shapes. It can be folded and flattened without signs of breaking or failure. SAE 6150 is used for making springs, while chrome-vanadium with high-carbon content, SAE 6195, is used for ball and roller bearings.

Molybdenum in small percentages is used in combination with chromium to form chrome-molybdenum steel, which has various uses in aircraft. Molybdenum is a strong alloying element that raises the ultimate strength of steel without affecting ductility or workability. Molybdenum steels are tough and wear resistant, and they harden throughout when heat treated. They are especially adaptable for welding and, for this reason, are used principally for welded structural parts and assemblies. This type of steel has practically replaced carbon steel in the fabrication of fuselage tubing, engine mounts, landing gears, and other structural parts. For example, a heat-treated SAE 4130 tube is approximately four times as strong as an SAE 1025 tube of the same weight and size.

A series of chrome-molybdenum steel most used in aircraft construction contains 0.25 to 0.55 percent carbon, 0.15 to 0.25 percent molybdenum, and 0.50 to 1.10 percent chromium. These steels, when suitably heat treated, are deep hardening, easily machined, readily welded by either gas or electric methods, and are especially adapted to high-temperature service.

Inconel is a nickel-chromium-iron alloy that closely resembles stainless steel in appearance. Because these two metals look very much alike, a distinguishing test is often necessary. One method of identification is to use a solution of 10 grams of cupric chloride in 100 cubic centimeters of hydrochloric acid. With a medicine dropper, place one drop of the solution on a sample of each metal to be tested and allow it to remain for two minutes. At the end of this period, slowly add three or four drops of water to the solution on the metal samples, one drop at a time; then wash the samples in clear water and dry them. If the metal is stainless steel, the copper in the cupric chloride solution will be deposited on the metal leaving a copper-colored spot. If the sample is inconel, a new-looking spot will be present.

The tensile strength of inconel is 100,000 psi annealed, and 125,000 psi, when hard rolled. It is highly resistant to salt water and is able to withstand temperatures as high as 1600°F. Inconel welds readily and has working qualities quite similar to those of corrosion-resistant steels.

Heat treatment of ferrous metals

The first important consideration in the heat treatment of a steel part is to know its chemical composition. This, in turn, determines its upper critical point. When the upper critical point is known, the next consideration is the rate of heating and cooling to be used. Carrying out these operations involves the use of uniform heating furnaces, proper temperature controls, and suitable quenching mediums.

Heat treating requires special techniques and equipment that are usually associated with manufacturers or large repair stations. Figure 3-21 shows the various heat-treating procedures for steel. The heat treatment of alloy steels includes hardening, tempering, annealing, normalizing, casehardening, carburizing, and nitriding.

Steel No.	Temperatures Normalizing air cool (°F)	Annealing (°F)	Hardening (°F)	Quenching medium (n)	Tempering (drawing) temperature for tensile strength (psi) 100,000 (°F)	125,000 (°F)	150,000 (°F)	180,000 (°F)	200,000 (°F)
1020	1,650–1,750	1,600–1,700	1,575–1,675	Water	—	—	—	—	—
1022 (x1020)	1,650–1,750	1,600–1,700	1,575–1,675	Water	—	—	—	—	—
1025	1,600–1,700	1,575–1,650	1,575–1,675	Water	(a)	—	—	—	—
1035	1,575–1,650	1,575–1,625	1,525–1,600	Water	875	—	—	—	—
1045	1,550–1,600	1,550–1,600	1,475–1,550	Oil or water	1,150	—	—	(n)	—
1095	1,475–1,550	1,450–1,500	1,425–1,500	Oil	(b)	—	1,100	850	750
2330	1,475–1,525	1,425–1,475	1,450–1,500	Oil or water	1,100	950	800	—	—
3135	1,600–1,650	1,500–1,550	1,475–1,525	Oil	1,250	1,050	900	750	650
3140	1,600–1,650	1,500–1,550	1,475–1,525	Oil	1,325	1,075	925	775	700
4037	1,600	1,525–1,575	1,525–1,575	Oil or water	1,225	1,100	975	—	—
4130 (x4130)	1,600–1,700	1,525–1,575	1,525–1,625	Oil (c)	(d)	1,050	900	700	575
4140	1,600–1,650	1,525–1,575	1,525–1,575	Oil	1,350	1,100	1,025	825	675
4150	1,550–1,600	1,475–1,525	1,550–1,550	Oil	—	1,275	1,175	1,050	950
4340 (x4340)	1,550–1,625	1,525–1,575	1,475–1,550	Oil	—	1,200	1,050	950	850
4640	1,675–1,700	1,525–1,575	1,500–1,550	Oil	—	1,200	1,050	750	625
6135	1,600–1,700	1,550–1,600	1,575–1,625	Oil	1,300	1,075	925	800	750
6150	1,600–1,650	1,525–1,575	1,550–1,625	Oil	(d)(e)	1,200	1,000	900	800
6195	1,600–1,650	1,525–1,575	1,500–1,550	Oil	(f)	—	—	—	—
NE8620	—	—	1,525–1,575	Oil	—	1,000	—	—	—
NE8630	1,650	1,525–1,575	1,525–1,575	Oil	—	1,125	975	775	675
NE8735	1,650	1,525–1,575	1,525–1,575	Oil	—	1,175	1,025	875	775
NE8740	1,625	1,500–1,550	1,500–1,550	Oil	—	1,200	1,075	925	850
30905	—	(g)(h)	(i)	—	—	—	—	—	—
51210	1,525–1,575	1,525–1,575	1,775–1,825 (j)	Oil	1,200	1,100	(k)	750	—
51335	—	1,525–1,575	1,775–1,850	Oil	—	—	—	—	—
52100	1,625–1,700	1,400–1,450	1,525–1,550	Oil	(f)	—	—	—	—
Corrosion resisting (16-2)(1)	—	—	—	—	(m)	—	—	—	—
Silicon Chromium (for springs)	—	—	1,700–1,725	Oil	—	—	—	—	—

Notes:

(a) Draw at 1,150°F for tensile strength of 70,000 psi.

(b) For spring temper draw at 800–900°F. Rockwell hardness C-40–45.

(c) Bars or forgings may be quenched in water from 1,500–1,600°F.

(d) Air cooling from the normalizing temperature will produce a tensile strength of approximately 90,000 psi.

(e) For spring temper draw at 850–950°F. Rockwell hardness C-40–45.

(f) Draw at 350–450°F to remove quenching strains. Rockwell hardness C-60–65.

(g) Anneal at 1,600–1,700°F to remove residual stresses due to welding or cold work. May be applied only to steel containing titanium or columbium.

(h) Anneal at 1,900–2,100°F to produce maximum softness and corrosion resistance. Cool in air or quench in water.

(i) Harden by cold work only.

(j) Lower side of range for sheet 0.06 inch and under. Middle of range for sheet and wire 0.125 inch. Upper side of range for forgings.

(k) Not recommended for intermediate tensile strengths because of low impact.

(l) AN-QQ-S-770 — It is recommended that, prior to tempering, corrosion-resisting (16 Cr-2 Ni) steel be quenched in oil from a temperature of 1,875–1,900°F, after a soaking period of 30 minutes at this temperature. To obtain a tensile strength at 115,000 psi, the tempering temperature should be approximately 525°F. A holding time at these temperatures of about 2 hours is recommended. Tempering temperatures between 700°F and 1,100°F will not be approved.

(m) Draw at approximately 800°F and cool in air for Rockwell hardness of C-50.

(n) Water used for quenching shall be within the temperature range of 80–150°F.

Figure 3-21 Heat-treatment procedures for steels.

4

Drilling and Countersinking

Although drilling holes seems a simple task, it requires a great deal of knowledge and skill to do it properly and in accordance with specifications. It is one of the most important operations performed by riveters or mechanics. With enough study and a considerable amount of practice, practically anyone can learn to perform the operation.

Rivet Hole Preparation

Preparing holes to specifications requires more than just running a drill through a piece of metal. This chapter outlines the fundamentals of preparing proper holes, primarily for all types of rivets and rivet-type fasteners; however, the information is also generally applicable to bolts, pins, or any other devices that require accurately drilled holes.

Countersinking is another phase of preparing holes for certain types of fasteners. Countersinking procedures and other related data are also included in this chapter.

Rivet hole location

Before drilling any hole, it is necessary to know where to drill it. This can be done by any one or a combination of the following methods:

- By pilot holes punched while the part is being made on a punch press and enlarging the holes to full size on assembly.

- By use of a template.
- By drilling through drill bushings in a jig on assembly.
- By using a "hole finder" to locate holes in the outer skin over the pilot or predrilled hole in the substructure.
- By laying out the rivet pattern by measurements from a blueprint. When it is necessary to mark hole locations, a colored pencil that contains no lead, or a water soluble fine point felt pen should be used. The carbon in lead pencils is highly incompatible with aluminum and should not be used. Never use a scriber or other similar object that would scratch the metal.

Drills

Rivet holes are generally made with an air drill motor and a standard straight shank twist drill, as shown in Fig. 4-1.

Twist drills for most aircraft work are available in three different size groups: "letter" sizes A through Z; "number" sizes 80 through 1; and "fractional" sizes, from diameters of $\frac{1}{64}$ inch up to $1\frac{1}{4}$ inch, increasing in increments of $\frac{1}{64}$ inch. "Fractional" sizes are also available in larger diameters, but are not used for rivet fasteners. All drill sizes are marked on the drill shank. See Fig. 4-2 for normally available drill sizes.

Drills are made from the following materials:

- *Carbon Steel* Not normally used in the aerospace industry because of its inferior working qualities to high-speed steel.
- *High-Speed Steel* Most drills used in the aerospace industry are high-speed steel because of good physical characteristics, ready availability, and because they do not present any difficult problems in resharpening.
- *Cobalt Alloy Steels* Used on high heat-treated steels over 180,000 psi.
- *Cemented Carbide Inserts* Used for cutting very hard and abrasive materials. Limited use in the aerospace industry.

Drill sizes are not always readable on the drill shank because the drill chuck has spun on the drill and removed the markings.

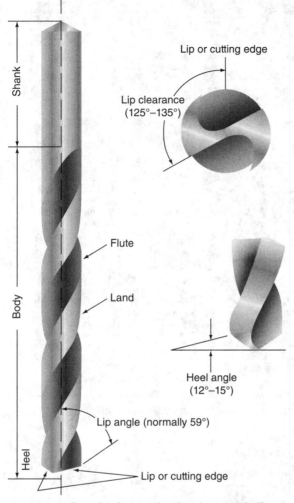

Figure 4-1 Standard straight shank twist drill.

If the drill size cannot easily be read on the drill, always use a drill gauge, shown in Fig. 4-3.

Drill sharpening

The twist drill should be sharpened at the first sign of dullness. Faulty sharpening accounts for most of the difficulty encountered

Drill Size	Decimal (Inches)	Drill Size	Decimal (Inches)	Drill Size	Decimal (Inches)	Drill Size	Decimal (Inches)	Drill Size	Decimal (Inches)
80	.0135	50	.0700	22	.1570	G	.2610	31/64	.4844
79	.0145	49	.0730	21	.1590	17/64	.2656	1/2	.5000
1/54	.0156	48	.0760	20	.1610	H	.2660	33/64	.5156
78	.0160	5/64	.0781	19	.1660	I	.2720	17/32	.5312
77	.0180	47	.0785	18	.1695	J	.2770	35/64	.5469
76	.0200	46	.0810	11/64	.1718	K	.2810	9/16	.5625
75	.0210	45	.0820	17	.1730	9/32	.2812	37/64	.5781
74	.0225	44	.0860	16	.1770	L	.2900	19/32	.5937
73	.0240	43	.0890	15	.1800	M	.2950	39/64	.6094
72	.0250	42	.0935	14	.1820	19/64	.2968	5/8	.6250
71	.0260	3/32	.0937	13	.1850	N	.3020	41/64	.6406
70	.0280	41	.0960	3/16	.1875	5/16	.3125	21/32	.6562
69	.0293	40	.0980	12	.1890	O	.3160	43/64	.6719
68	.0310	39	.0995	11	.1910	P	.3230	11/16	.6875
1/32	.0312	38	.1015	10	.1935	21/64	.3281	45/64	.7031
67	.0320	37	.1040	9	.1960	Q	.3320	23/32	.7187
66	.0330	36	.1065	8	.1990	R	.3390	47/64	.7344
65	.0350	7/64	.1093	7	.2010	11/32	.3437	3/4	.7500
64	.0360	35	.1100	13/64	.2031	S	.3480	49/64	.7656
63	.0370	34	.1110	6	.2040	T	.3580	25/32	.7812
62	.0380	33	.1130	5	.2055	23/64	.3593	51/64	.7969
61	.0390	32	.1160	4	.2090	U	.3680	13/16	.8125
60	.0400	31	.1200	3	.2130	3/8	.3750	53/64	.8281
59	.0410	1/8	.1250	7/32	.2187	V	.3770	27/32	.8437
58	.0420	30	.1285	2	.2210	W	.3860	55/64	.8594
57	.0430	29	.1360	1	.2280	25/64	.3906	7/8	.8750
56	.0465	28	.1405	A	.2340	X	.3970	57/64	.8906
3/64	.0468	9/64	.1406	15/64	.2343	Y	.4040	29/32	.9062
55	.0520	27	.1440	B	.2380	13/32	.4062	59/64	.9219
54	.0550	26	.1470	C	.2420	Z	.4130	15/16	.9375
53	.0595	25	.1495	D	.2460	27/64	.4219	61/64	.9531
1/16	.0625	24	.1520	1/4	.2500	7/16	.4375	31/32	.9687
52	.0635	23	.1540	E	.2500	29/64	.4531	63/64	.9844
51	.0670	5/32	.1562	F	.2570	15/32	.4687	1	1.0000

Figure 4-2   Sizes and designations of fraction, number, and letter drills.

Figure 4-3   Drill gauges: fractions on the left and number on the right. Decimal equivalents are also given.

Courtesy L.S. Starret Company

Figure 4-4 Drill-sharpening gauge.
Courtesy L.S. Starret Company

in drilling. Although drills can be sharpened by hand, a drill-sharpening jig should be used when available. Using the drill gauge (Fig. 4-4), rotating the drill about its central axis will not provide the 12-degree lip clearance required. The drill must be handled so that the heel will be ground lower than the lip. Using the drill gauge, it is possible to maintain equal length lips that form equal angles with the central axis. If the drill is rotated slightly the gauge will indicate whether the heel has sufficient clearance.

Typical procedures for sharpening drills are as follows (see Fig. 4-5):

1. Adjust the grinder tool rest to a convenient height for resting the back of the hand while grinding.
2. Hold the drill between the thumb and index finger of the right or left hand. Grasp the body of the drill near the shank with the other hand.
3. Place the hand on the tool rest with the centerline of the drill making a 59-degree angle with the cutting face of the grinding wheel. Lower the shank end of the drill slightly.
4. Slowly place the cutting edge of the drill against the grinding wheel. Gradually lower the shank of the drill as you twist the drill in a clockwise direction. Maintain pressure against the grinding surface only until you reach the heel of the drill.
5. Check the results of grinding with a gauge to determine whether or not the lips are the same length and at a 59° or other desired angle.

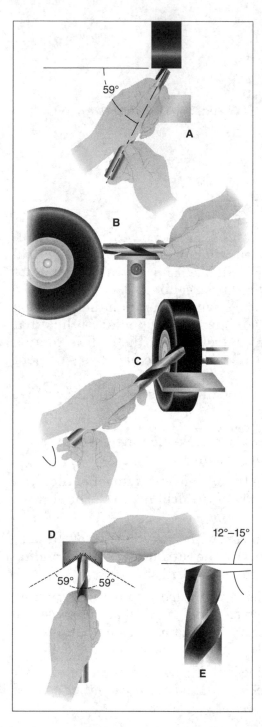

Figure 4-5 Drill-sharpening procedure.

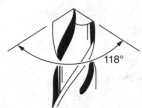
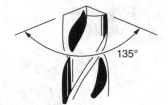

GENERAL PURPOSE
ALUMINUM, MAGNESIUM
MILD STEEL

HARD AND TOUGH MATERIALS
STAINLESS STEEL, HARD
STEEL, TITANIUM

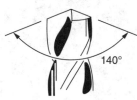

COBALT DRILLS FOR HIGH
HEAT TREAT STEELS

PLEXIGLASS AND KIRKSITE
ALSO USED FOR ENLARGING
HOLES IN THIN SHEET

Figure 4-6 Typical drill points for drilling various materials.

Drill points

Drills are made with a number of different points or are ground to different angles for a specific application, as shown in Fig. 4-6. Always select the correct shape point for the job. As a general rule, the point angle should be flat or large for hard and tough materials, and sharp or small for soft materials.

A 135-degree split point drill point is most often used to drill heat-treated aluminum alloys. The split point reduces the tendency of "drill walking" during the start of the drilling operation.

Drilling equipment

The air drill motor is used in the aerospace industry in preference to an electric motor because the air motor has no fire or shock hazards, has a lower initial cost, requires less maintenance, and running speed is easier to control. Air motors are available in a variety of sizes, shapes, running speeds, and drilling head angles (Fig. 4-7).

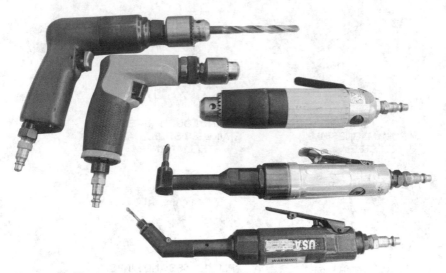

Figure 4-7 Typical air motors used for sheet metal repair.

Drilling Operations

Chucking the drill

WARNING

Before installing or removing drill bits, countersinks, or other devices in an air motor, be sure that the air line to the motor is disconnected. Failure to observe this precaution can cause serious injury.

1. **Install proper drill in the motor and tighten with proper size chuck key. Be sure to center the drill in the chuck. Do not allow flutes to enter the chuck.**
2. **Connect the air hose to the motor inlet fitting.**
3. **Start the drill motor and check the drill for wobble. The drill must run true, or an oversize hole will be made. Replace bent drills.**

Drilling holes

1. Hold the motor firmly. Hold the drill at 90-degree angle to the surface, as shown in Fig. 4-8.

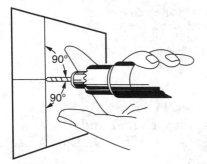

Figure 4-8 Square drill with work.

2. Start the hole by placing the point of the drill on the marked centerline. With the fingers, turn the chuck until an indentation is made. (Omit this step when drilling through a drill bushing or when a pilot hole exists.)

3. Position thumb and forefinger to prevent the drill from going too far through the work, which can cause damage to items on the other side or result in an oversized hole.

4. Drill the hole by starting the drill motor and exerting pressure on the centerline of the drill. Exert just enough pressure to start the drill cutting a fairly large size of chip and maintain this pressure until the drill starts to come through the work.

5. Decrease the pressure and cushion the breakthrough with the fingers when the drill comes through. Do not let the drill go any farther through the hole than is necessary to make a good, clean hole. Do not let the drill spin in the hole any longer than necessary.

6. Withdraw the drill from the hole in a straight line perpendicular to the work. Keep motor running while withdrawing drill.

To ensure proper centering and a correct, final-sized hole, rivet holes are usually pilot drilled with a drill bit that is smaller than the one used to finish the hole. Selected larger-diameter holes must be predrilled after pilot drilling and before final-sized drilling to ensure a round, accurate hole for the rivet. This procedure is sometimes referred to as *step drilling*.

Note: When drilling thin sheet-metal parts, support the part from the rear with a wooden block or other suitable material to prevent bending.

Figure 4-9 For cutting clean, large-diameter holes in thin sheet metal, hole saws are commonly used.

For enlarging holes in thin sheet metal use:

Plastic-type drills for hole diameter ⅜ inch and under.

Hole saws for holes over ⅜-inch diameter (Fig. 4-9). Do not use counterbores or spotfacers.

Drill stops and drill bushings

Always use a drill stop shown in Fig. 4-10 to prevent excessive drill penetration that might damage underlying structure, injure personnel, and prevent the drill chuck from marring the surface. Drill bushings as shown in Fig. 4-11 are used to hold the drill perpendicular to the part and prevent also excessive drill penetration and possible damage to the aircraft skin.

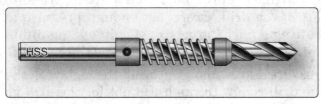

Figure 4-10 Drill stop.

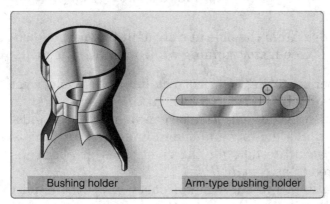

Bushing holder   Arm-type bushing holder

Figure 4-11 Drill bushings.

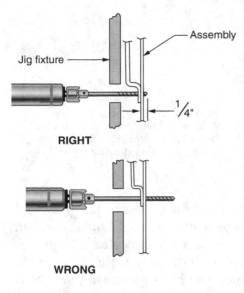

Jig fixture

Assembly

$\frac{1}{4}$"

RIGHT

WRONG

Figure 4-12   Select a drill of
the correct length and size.

Using an extension drill

Special drills can be used with the air-drill motor. The long drill
(sometimes called a *flexible drill*) comes in common drill sizes and
in 6-inch, 8-inch, 10-inch, or longer lengths. Do not use a longer
drill than necessary. See Fig. 4-12.

CAUTION

1. **Before starting the motor, hold the extension near the
   flute end with one hand as shown in Fig. 4-13. Don't
   touch the flutes and don't forget to wear safety glasses
   or a face shield.**
2. **Drill through the part. Do not let go of the drill shank.
   Keep the motor running as the drill is removed.**

Drilling aluminum and aluminum alloys

Drilling these materials has become quite commonplace and few
difficulties are experienced. Some of the newer aluminum alloys
of high silicon content and some of the cast alloys still present
several problems.

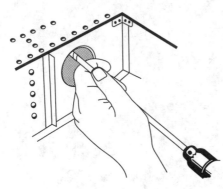

Figure 4-13 Hold the extension drill near the flute end with one hand. An unsupported drill might whip around and cause injury.

General-purpose drills can be used for all sheet material.

High rates of penetration can be used when drilling aluminum; hence, disposal of chips or cuttings is very important. To permit these high-penetration rates and still dispose of the chips, drills have to be free cutting to reduce the heat generated and have large flute areas for the passage of chips.

Although the mechanic has no direct indication of drill motor speed, a relatively high rpm can be used.

Drilling titanium and titanium alloys

Titanium and its alloys have low-volume specific heat and low thermal conductivity, causing them to heat readily at the point of cutting, and making them difficult to cool because the heat does not dissipate readily.

Thermal problems can best be overcome by reducing either the speed or the feed. Fortunately, titanium alloys do not work-harden appreciably, thus lighter feed pressures can be used.

When using super-high-speed drills containing high carbon, vanadium, and cobalt to resist abrasion and high drilling heats, a speed (rpm) considerably slower than for aluminum must be used.

See Chap. 3 for further information regarding the drilling of titanium.

Drilling stainless steel

Stainless steel is more difficult to drill than aluminum alloys and straight carbon steel because of the work-hardening properties. Because of work hardening, it is most important to cut continuously with a uniform speed and feed. If the tool is permitted to rub or idle on the work, the surface will become work hardened to a point where it is difficult to restart the cut.

For best results in cutting stainless steel, the following should be adhered to:

- Use sharp drills, point angle 135 degrees.

- Use moderate speeds.

- Use adequate and uniform feeds.

- Use an adequate amount of sulfurized mineral oil or soluble oil as a coolant, if possible.

- Use drill motor speeds the same as for titanium.

Hint. When drilling through dissimilar materials, drill through the harder material first to prevent making an egg-shaped hole in the softer material.

Deburring

Drilling operations cause burrs to form on each side of the sheet and between sheets. Removal of these burrs, called *debunking* or *burring*, must be performed if the burrs tend to cause a separation between the parts being riveted. Burrs under either head of a rivet do not, in general, result in unacceptable riveting. The burrs do not have to be removed if the material is to be used immediately; however, sharp burrs must be removed, if the material is to be stored or stacked, to prevent scratching of adjacent parts or injury to personnel.

Care must be taken to limit the amount of metal removed when burrs are removed. Removal of any appreciable amount of material from the edge of the rivet hole will result in a riveted joint of lowered strength. Deburring shall not be performed on predrilled holes that are to be subsequently form countersunk.

Remove drill chips and dirt prior to riveting to prevent separation of the sheets being riveted. Burrs and chips can be minimized by clamping the sheets securely during drilling and backing up the work if the rear member is not sufficiently rigid. A "chip chaser" (Fig. 4-14) can be used when necessary to remove loose chips between the material.

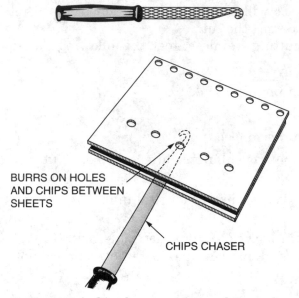

BURRS ON HOLES
AND CHIPS BETWEEN
SHEETS

CHIPS CHASER

Figure 4-14   A chip chaser can be used to remove chips between material.

Countersinking

Flush head rivets (100° countersunk) require a countersunk hole prepared for the manufactured rivet head to nest in. This is accomplished by one of two methods: machine countersinking or form countersinking (dimpling), as shown in Fig. 4-15.

SURFACE COUNTERSUNK

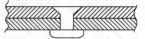

SURFACE DIMPLED

SUB-SURFACE COUNTERSUNK

SURFACE DIMPLED

SUB-SURFACE DIMPLED

Figure 4-15   Countersinking and dimpling.

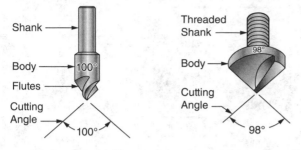

Figure 4-16 Straight shank and rosebud counter-sinking cutters.

Types of countersinking cutters

The straight shank cutter is shown in Fig. 4-16. The cutting angle is marked on the body. Cutting angles commonly used are 100 and 110 degrees. The diameter of the body varies from ¼ to 1½ inch. A countersink of ⅜-inch diameter is most commonly used.

A countersink cutter (rose bud) for angle drills, also shown in Fig. 4-16, is used if no other countersink will do the job.

The stop countersink (Fig. 4-17) consists of the cutter and a cage. The cutter has a threaded shank to fit the cage and an integral pilot. The cutting angle is marked on the body. The cage consists of a foot piece, locking sleeve, locknut, and spindle. The foot-piece is also available in various shapes and sizes. Stop countersinks must be used in all countersinking operations, except where there is not enough clearance.

CAUTION

When using a stop countersink, always hold the skirt firmly with one hand. If the countersink turns or vibrates, the material will be marred and a ring will be made around the hole.

Back (inserted) countersinks (Fig. 4-18) should be used when access for countersinking is difficult. The back countersink consists of two pieces: a rod, of the same diameter as the drilled hole, which slips through the hole, and a cutter that is attached on the far side.

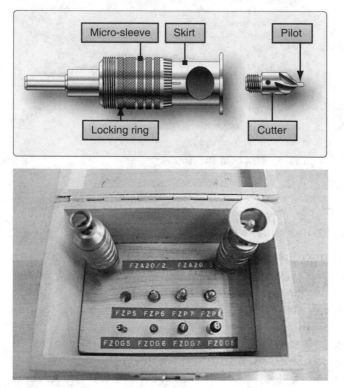

Figure 4-17 The stop countersink set with go no-go gages.

Countersinking holes

To countersink holes, proceed as follows:

1. Inspect the holes to be countersunk. The holes must be of the proper size, perpendicular to the work surface, and not be elongated.

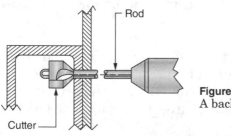

Figure 4-18
A back countersink.

2. Select the proper size of countersink. The pilot should just fit the hole and turn freely in the hole. If the hole is too tight, the cutter will "freeze-up" in the hole and might break.

3. Check the angle of the countersink.

4. Set the depth of the stop countersink on a piece of scrap before countersinking a part. Always check for proper head flushness by driving a few rivets of the required type and size in the scrap material. The rivet heads should be flush after driving. In some cases, where aerodynamic smoothness is a necessity, the blueprint might specify that countersunk holes be made so that flush head fasteners will be a few thousandths of an inch high. Such fasteners are shaved to close limits after driving.

5. Countersink the part. Be sure to hold the skirt to keep it from marking the part and apply a steady pressure to the motor to keep the cutter from chattering in the hole.

Minimum countersinking depth

If countersinking is done on metal below a certain thickness, a knife edge with less than the minimum bearing surface or actual enlarging of the hole may result. The general rule for countersinking and flush fastener installation procedures has been reevaluated in recent years because countersunk holes have been responsible for fatigue cracks in aircraft pressurized skin. In the past, the general rule for countersinking held that the fastener head must be contained within the outer sheet. A combination of countersinks too deep (creating a knife edge), number of pressurization cycles, fatigue, deterioration of bonding materials, and working fasteners caused a high stress concentration that resulted in skin cracks and fastener failures. Some manufacturers are currently recommending the countersink depth be no more than $\frac{2}{3}$ the outer sheet thickness or down to 0.020-inch minimum fastener shank depth, whichever is greater. Dimple the skin if it is too thin for machine countersinking. Figure 4-19 shows the countersink dimension.

Form countersinking (dimpling)

Blueprints often specify form countersinking to form a stronger joint than machine countersinking provides. The sheet is not

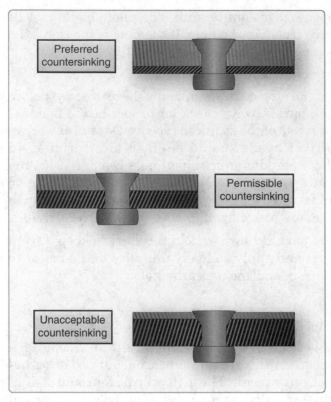

Figure 4-19 Countersinking dimensions.

weakened by cutting metal away, but is formed to interlock with the substructure. The two types of form countersinking accepted are coin dimpling and modified radius dimpling.

If flush rivets are required but the aircraft skin is too thin to be machine countersunk, the skin must be dimpled.

Coin dimpling. Coin dimpling is accomplished by using either a portable or a stationary squeezer, fitted with special dimpling dies (Fig. 4-20). These special dies consist of a male die held in one jaw of the squeezer and a female die held in the other jaw. In the female die, a movable coining ram exerts controlled pressure on the underside of a hole, while the male die exerts controlled pressure on the upper side to form a dimple. Pressure applied by the coining ram forms, or "coins," a dimple in the exact shape of the dies. Coin dimpling does not bend or stretch the material, as did the now-obsolete

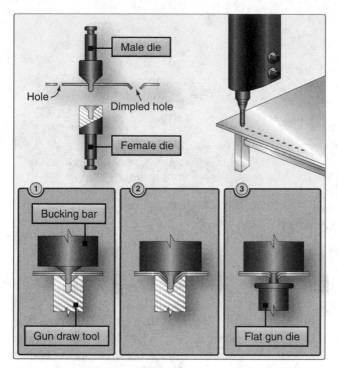

Figure 4-20   Dimpling techniques.

radius-dimpling system, and the dimple definition is almost as sharp as that of a machine countersink. Because the lower and upper sides of the dimple are parallel, any number of coined dimples can be nested together or into a machine countersink and the action of the coining ram prevents cracking of the dimple.

Coin dimpling is used on all skins when form countersinking is specified, and, wherever possible, on the substructure. When it is impossible to get coin-dimpling equipment into difficult places on the substructure, a modified radius dimple can be used and a coin dimple can then nest in another coin dimple, or a machine countersink, or a modified radius dimple. Unless the drawing specifies otherwise, dimpling shall be performed only on a single thickness of material.

Modified radius dimpling.   The modified radius dimple is similar to the coin dimple, except that the coining ram is stationary in the female die and is located at the bottom of the recess (Fig. 4-21). Because the pressure applied by the stationary coining ram

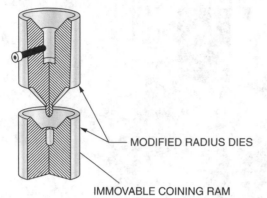

MODIFIED RADIUS DIES

IMMOVABLE COINING RAM

Figure 4-21 Modified-radius dimple dies.

cannot be controlled, the amount of forging or coining is limited. The modified radius dimple does not have as sharp a definition as the coin dimple. Because the upper and lower sides of the modified radius dimple are not parallel, this type of dimple can never nest into another dimple or countersink, and when used must always be the bottom dimple. The advantage of the modified radius dimple is that the dimpling equipment can be made smaller and can get into otherwise inaccessible places on the substructure. Dimples for panel fasteners, such as Dzus, Camloc, and Airloc fasteners, might be modified radius dimpled.

Heat is used with some types of material when doing either type of form dimpling. Magnesium, titanium, and certain aluminum alloys must be dimpled with heated dies. Primed surfaces can be hot or cold dimpled, depending on the metal, and heat can be used to dimple any material, except stainless steel, to prevent cracking. A ram coin hot dimpler is shown in Fig. 4-22.

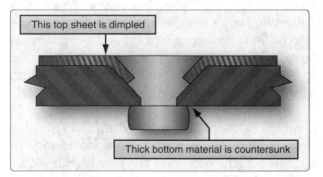

This top sheet is dimpled

Thick bottom material is countersunk

Figure 4-22 100° combination predimple and countersink method.

100° combination predimple and countersink method

Metals of different thicknesses are sometimes joined by a combination of dimpling and countersinking. See Fig. 4-22. A countersink well made to receive a dimple is called a subcountersink. These are most often seen where a thin web is attached to heavy structure. It is also used on thin gap seals, wear strips, and repairs for worn countersinks.

Hole preparation for form countersinking

Preparation of holes for form countersinking is of great importance because improperly drilled holes result in defective dimples. Holes for solid-shank rivets must be size drilled, before dimpling, by using the size drills recommended for regular holes. Holes for other fasteners must be predrilled before dimpling, and then drilled to size, according to the blueprint or applicable specification after dimpling. Do not burr holes to be form countersunk, except on titanium.

CAUTION

Form countersinking equipment (coin dimpling and modified radius dimpling) is normally operated only by certified operators who have been instructed and certified to operate this equipment.
To accomplish general dimpling, proceed as follows:

1. Fit skin in place on substructure.
2. Pilot drill all holes (Cleco often).
3. Drill to proper size for dimpling: final size for conventional rivets; predrill size for all other rivets.
4. Mark all holes according to NAS523 rivet code letters (see Chap. 10) to show the type and size of fastener before removing the skin or other parts from the assembly. Mark "DD," which means *dimple down*, with a grease pencil on the head side of the part.
5. Remove the skin and have it dimpled.
6. Have the substructure dimpled or countersunk as specified on the blueprint. Mark it, as in step 4.

7. Size drill holes when necessary.

8. Fit the skin.

9. Install the rivets.

Shaving Flush Head Fasteners

Rivets, bolts, screws, or other fasteners that protrude above the surface (beyond allowable tolerances for aerodynamic smoothness) might require shaving. The amount that a rivet can protrude above the surface of the skin varies with each airplane model and with different surfaces on the airplane. Rivet shaving (milling) is accomplished with an air-driven, high-speed cutter in a rivet shaver, as shown in Fig. 4-23.

After shaving, fasteners should be flush within 0.001 inch above the surface—even though a greater protuberance is allowable in that particular area for unshaved fasteners.

WARNING
Shaved fasteners have a sharp edge and could be a hazard to personnel.

Shaved rivets and abraded areas adjacent to shaved rivets and blind rivets that have broken pin ends and are located in parts,

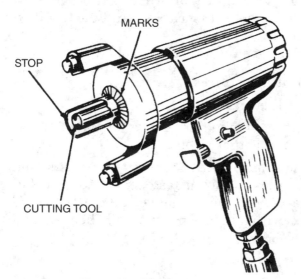

Figure 4-23 Typical rivet shaver.

for which applicable drawings specify paint protection, must be treated for improved paint adhesion.

Reamers

Reamers are used to smooth and enlarge holes to the exact size. Hand reamers have square end shanks so that they can be turned with a tap wrench or a similar handle. Various reamers are illustrated in Fig. 4-24.

A hole that is to be reamed to exact size must be drilled about 0.003- to 0.007-inch undersize. A cut that removes more than 0.007 inch places too much load on the reamer and should not be attempted.

Reamers are made of either carbon tool steel or high-speed steel. The cutting blades of a high-speed steel reamer lose their original keenness sooner than those of a carbon steel reamer;

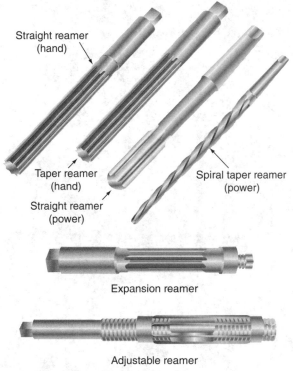

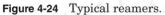

Figure 4-24 Typical reamers.

however, after the first superkeenness is gone, they are still serviceable. The high-speed reamer usually lasts much longer than the carbon steel type.

Reamer blades are hardened to the point of being brittle and must be handled carefully to avoid chipping them. When reaming a hole, rotate the reamer in the cutting direction only. Turn the reamer steadily and evenly to prevent chattering, marking, and scoring the hole area.

Reamers are available in any standard size. The straight-fluted reamer is less expensive than the spiral-fluted reamer, but the spiral type has less tendency to chatter. Both types are tapered for a short distance back of the end to aid in starting. Bottoming reamers have no taper and are used to complete the reaming of blind holes.

For general use, an expansion reamer is the most practical. This type is furnished in standard sizes from $\frac{1}{4}$ to 1 inch, increasing in diameter by $\frac{1}{32}$ inch increments. Taper reamers, both hand- and machine-operated, are used to smooth and true tapered holes and recesses.

Riveting

Riveting is the strongest practical means of fastening airplane skins and the substructure together. Although the cost of installing one rivet is small, the great number of rivets used in airplane manufacture represents a large percentage of the total cost of any airplane.

Solid-Shank Rivets

Although many special rivets are covered later in this chapter, solid-shank (conventional) rivets are the most commonly used rivets in aircraft construction. They consist of a manufactured head, a shank, and a driven head. The driven head, sometimes called a *shop head* or *upset head*, is caused by upsetting the shank with a rivet gun or rivet squeezer. The shank actually expands slightly while being driven so the rivet fits tightly in the drilled hole (Fig. 5-1).

Material

Solid-shank rivets are manufactured from several kinds of metal or different alloys of these metals to fulfill specific requirements. These different metals and alloys are usually specified in a rivet designation by a system of letters. They are further identified by a system of markings on the rivet head. In some cases, the absence of a head marking signifies the alloy within a particular alloy group, or a particular color is used for a

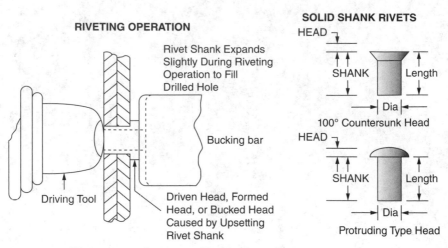

Figure 5-1   Rivet nomenclature and basic operation.

particular alloy. Figure 5-2 shows the more commonly used aluminum alloy rivets.

Rivet types and identification

In the past, solid-shank rivets with several different types of heads were manufactured for use on aircraft; now only two basic head types are used: countersunk and universal. However, in special cases, there are a few exceptions to this rule (Fig. 5-3).

Briles rivets (modified 120-degree countersunk rivet).   The Briles rivet as shown in Fig. 5-4 has a modified 120-degree head and is being used on several new aircraft. The Briles rivets are available in aluminum and titanium/columbium alloys (D - no mark on rivet head, KE - impressed ring on rivet head). The rivet holes can be drilled with standard drill bit sizes, but a special piloted counterbore/countersink must be used, due to the modified 120-degree head of the Briles rivet. Flushness can be checked by using a "dual gage." The Briles rivet can be driven with standard pneumatic rivet guns using standard "mushroom" rivet sets, or hand or pneumatic rivet squeezers. The rivet head expands simultaneously while upsetting the buck-tail. A major advantage of this rivet is that it can be driven in the heat-treated condition just like an AD rivet.

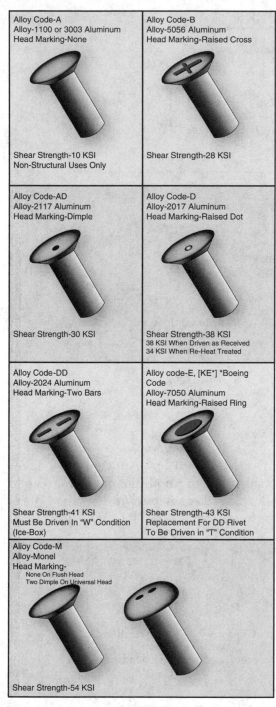

Figure 5-2 Rivet identification.

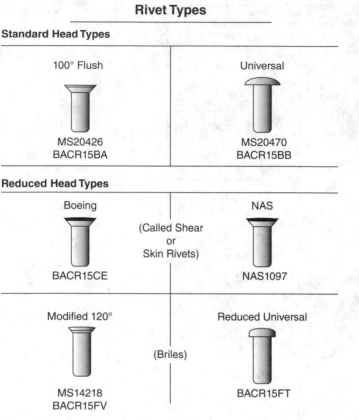

Figure 5-3 Common rivet head styles.

Reduced universal head rivet (BACR15FT). Another new rivet development is the reduced head or modified universal head rivets. These rivets were developed for use in areas where aerodynamic smoothness is unimportant. They look like universal

Figure 5-4 Briles modified 120 countersunk-counterbore rivet and reduced head universal rivet.

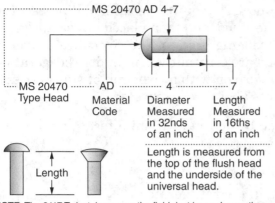

MS 20470 AD 4–7

MS 20470	AD	4	7
Type Head	Material Code	Diameter Measured in 32nds of an inch	Length Measured in 16ths of an inch

Length is measured from the top of the flush head and the underside of the universal head.

Length

NOTE: The 2117-T rivet, known as the field rivet is used more than any other for riveting aluminum alloy structures. The field rivet is in wide demand because it is ready for use as received and needs no further heat-treating or annealong. It also has a high resistance to corrosion.

Figure 5.5 Code breakdown.

head rivets with the head diameter reduced. By reducing the outer diameter, a small weight savings is realized. Small reductions like this add up to many pounds and lighter aircraft.

Rivets are identified by their MS (Military Standard) number, which superseded the old AN (Army-Navy) number. Both designations are still in use, however (Fig. 5-5).

The 2017-T and 2024-T rivets (Fig. 5-6) are used in aluminum alloy structures, where more strength is needed than is obtainable with the same size of 2117-T rivet. These rivets are annealed and must be kept refrigerated until they are to be driven. The 2017-T rivet should be driven within approximately one hour and the 2024-T rivet within 10 to 20 minutes after removal from refrigeration (Fig. 5-6).

Figure 5.6 "Icebox" rivets: Type D, 2017-T (left) and Type DD, 2024-T (right).

Raised tit

Two bars

These rivets, type D and DD, require special handling because they are heat treated, quenched, and then placed under refrigeration to delay the age-hardening process. The rivets are delivered to the shop as needed and are constantly kept under refrigeration until just before they are driven with a rivet gun or squeezer set.

Remember these points about icebox rivets:

- Take no more than can be driven in 15 minutes.
- Keep rivets cold with dry ice.
- Hit them hard, not often.
- Never put rivets back in the refrigerator.
- Put unused rivets in the special container provided.

SAFETY PRECAUTION

Dry ice has a temperature of –105°F. Handle carefully; it can cause a severe burn.

The 5056 rivet is used to rivet magnesium alloy structures because of its corrosion-resistant qualities in combination with magnesium.

E rivets made from 7050 aluminum alloy are a replacement for DD icebox rivets. They can be driven in the T condition and don't need to be refrigerated like DD rivets.

Riveting Practice

Edge distance

Edge distance is the distance from the edge of the material to the center of the nearest rivet hole (Fig. 5-7). If the drawing does not specify a minimum edge distance, also called edge margin for

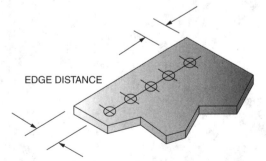

EDGE DISTANCE

Figure 5-7　Illustration of edge distance.

Edge distance/Edge Margin	Minimum Edge Distance	Preferred Edge Distance
Protruding head rivets	2 D	2 D + 1⅟₁₆"
Countersunk rivets	2½ D	2½ D + 1⅟₁₆"

Figure 5-8 Minimum edge distance/edge margin.

protruding head rivets (universal rivet), it is two times the diameter of the rivet shank, and for flush rivets the edge distance is 2½ times the rivet shank as shown in Fig. 5-8.

Rivet length

Solid-shank rivet lengths are never designated on the blueprint; the mechanic must select the proper length (Fig. 5-9).

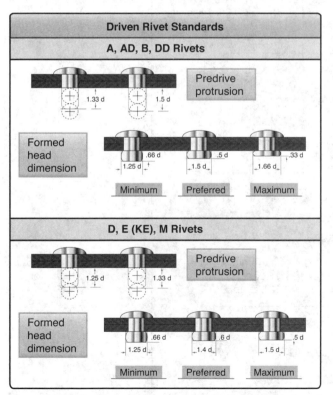

Figure 5-9 Predrive protrusion and formed head dimensions.

The standard rule for AD rivets is that the predriven length of the rivet is 1.5 times the diameter of the rivet. However, the predriven and formed head dimensions depend on the rivet material. The harder materials D, E, and M have a little shorter predriven length than the softer A, AD, B, and DD rivets.

Rivet spacing

Rivet spacing consists of rivet pitch and transfer pitch. Rivet pitch is the distance between the centers of neighboring rivets in the same row. Typical rivet spacing is between 4- and 6-rivet diameters (4D-6D). One-and three-row layouts have a minimum pitch of 3-rivet diameters, a two-row layout has a minimum pitch of 4-rivet diameters as shown in Figs. 5-10 and 5-11. The pitch

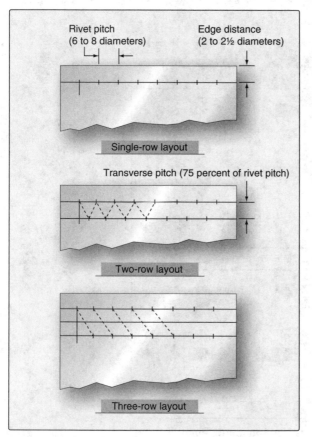

Figure 5-10 Rivet layout.

Rivet Spacing	Minimum Spacing	Preferred Spacing
1 and 3 rows protruding head rivet layout	3D	3D + 1/16"
2 row protruding head rivet layout	4D	4D + 1/16"
1 and 3 rows countersunk head rivet layout	3/1/2D	3/1/2D + 1/16"
2 row countersunk head rivet layout	4/1/2D	4/1/2D + 1/16"

Figure 5-11 Minimum and preferred rivet spacing.

for countersunk rivets is larger than for universal head rivets. Transverse pitch is the perpendicular distance between rivet rows. It is usually 75 percent of the rivet pitch. The smallest allowable transverse pitch is 2.5 rivet diameters.

Hole preparation

Consult Chap. 4 for hole-preparation details and for information on countersinking the holes and shaving of flush-head rivets. Drill sizes for various rivet diameters are shown in Fig. 5-12. Holes must be clean, round, and of the proper size. Forcing a rivet into a small hole will usually cause a burr to form under the rivet head.

Use of clecos

A cleco is a spring-loaded clamp used to hold parts together for riveting. Special pliers are used to insert clecos into holes (Fig. 5-13).

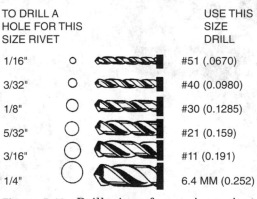

TO DRILL A
HOLE FOR THIS
SIZE RIVET

USE THIS
SIZE
DRILL

1/16"		#51 (.0670)
3/32"		#40 (0.0980)
1/8"		#30 (0.1285)
5/32"		#21 (0.159)
3/16"		#11 (0.191)
1/4"		6.4 MM (0.252)

Figure 5-12 Drill sizes for various rivet diameters.

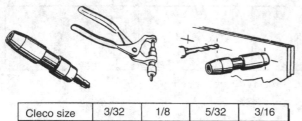

Cleco size	3/32	1/8	5/32	3/16
Color	Silver	Copper	Black	Brass

Figure 5-13 Clecos are inserted into holes with special cleco pliers. Cleco sizes are identified by colors.

Driving solid-shank rivets

Solid rivets can sometimes be driven and bucked by one operator using the conventional gun and bucking bar method when there is easy access to both sides of the work. In most cases, however, two operators are required to drive conventional solid-shank rivets.

Rivet guns.   Rivet guns vary in size, type of handle, number of strokes per minute, provisions for regulating speed, and a few other features. But, in general operation, they are all basically the same (Fig. 5-14). The mechanic should use a rivet-gun size that best suits the size of the rivet being driven. Avoid using too light a rivet gun because the driven head should be upset with the fewest blows possible.

Figure 5-14 Standard 3X and recoilless rivet gun.

Figure 5-15   Rivet sets.

NOTE
Always select a rivet gun size and bucking bar weight that will drive the rivet with as few blows as possible.

Rivet sets

Rivet sets (Fig. 5-15) are steel shafts that are inserted into the barrel of the rivet gun to transfer the vibrating power from the gun to the rivet head (Fig. 5-16).

Figure 5-16   The rivet gun and set go together like this.

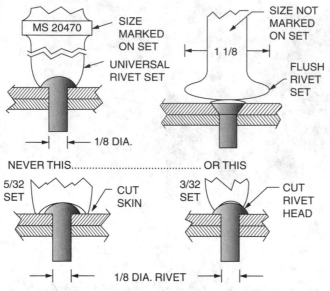

Figure 5-17   The correct set must be used for the rivet being driven.

Select a rivet set for the style of head and size of the rivet. Universal rivet sets can be identified with the tool number and size of the rivet. Flush sets can be identified only with the tool number (Fig. 5-17). Also shown in Fig. 5-17 is the result of using incorrect sets.

Bucking bars

A bucking bar is a piece of steel used to upset the driven head of the rivet. Bucking bars are made in a variety of sizes and shapes to fit various situations. Bucking bars must be handled carefully to prevent marring surfaces. When choosing a bucking bar to get into small places, choose one in which the center of gravity falls as near as possible over the rivet shank. Avoid using too light of a bucking bar because this causes excessive deflection of the material being riveted that, in turn, might cause marking of the outer skin by the rivet set. A bucking bar that is too heavy will cause a flattened driven head and might cause a loose manufactured head. Remember, you should use as heavy a bar as possible to drive the rivet with as few blows as possible. Figure 5-18 shows some typical bucking bar shapes.

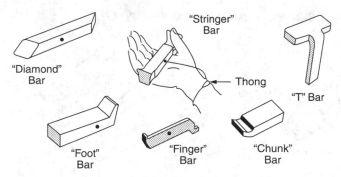

Figure 5-18 Some typical bucking bars.

Riveting procedure

Operate a rivet gun and install rivets as follows:

1. Install the proper rivet set in gun and attach the rivet set retaining spring, if possible. Certain flush sets have no provision for a retaining spring.

2. Connect the air hose to the gun.

3. Adjust the air regulator (Fig. 5-19), which controls the pressure or hitting power of the rivet gun, by holding the rivet set against a block of wood while pulling the trigger, which controls the operating time of the gun. The operator should time the gun to form the head in one "burst," if possible.

4. Insert proper rivet in hole.

5. Hold or wait for the bucker to hold the bucking bar on the shank of the rivet. The gun operator should "feel" the pressure being applied by the bucker and try to equalize this pressure.

6. Pull the gun trigger to release a short burst of blows. The rivet should now be properly driven, if the timing was correct, and provided that the bucking bar and gun were held firmly and perpendicular (square) with the work (Fig. 5-20).

Rivet gun operators should always be familiar with the type of structure beneath the skin being riveted and must realize the problems of the bucker (Fig. 5-21).

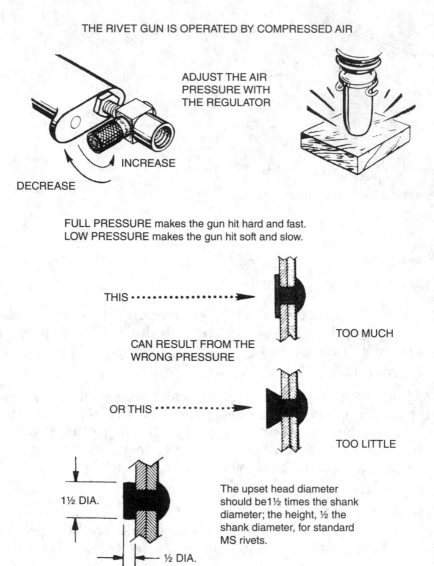

THE RIVET GUN IS OPERATED BY COMPRESSED AIR

ADJUST THE AIR PRESSURE WITH THE REGULATOR

INCREASE

DECREASE

FULL PRESSURE makes the gun hit hard and fast.
LOW PRESSURE makes the gun hit soft and slow.

THIS

CAN RESULT FROM THE WRONG PRESSURE

TOO MUCH

OR THIS

TOO LITTLE

1½ DIA.

½ DIA.

The upset head diameter should be 1½ times the shank diameter; the height, ½ the shank diameter, for standard MS rivets.

Figure 5-19 Adjust the air regulator that controls the hitting power of the gun by holding the rivet set against a block of wood.

CAUTION

Never operate a rivet gun on a rivet, unless it is being bucked. The bucker should always wait for the gun operator to stop before getting off a rivet.

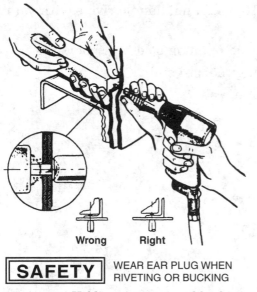

Figure 5-20 Holding rivet gun and bucking bar on rivet.

Skilled riveters:

- Use a slow action gun; it's easier to control.
- Use a 1⅛-inch bell-type rivet set for general-purpose flush riveting.
- Adjust the air pressure sufficiently to drive a rivet in two or three seconds.
- Use your body weight to hold the rivet gun and set firmly against the rivet.
- Hold the gun barrel at a 90-degree angle to the material.
- Squeeze the trigger by gripping it with your entire hand, as though you were squeezing a sponge rubber ball. Be sure that the bucking bar is on the rivet.

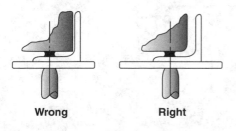

Figure 5-21 The bucker should not let the sharp corner of a bucking bar contact an inside radius of the skin or any other object.

- Operate the rivet gun with one hand; handle rivets with your other hand.
- Spot rivet the assembly; avoid reaming holes for spot rivets.
- Plan a sequence for riveting the assembly.
- Drive the rivets to a rhythm.

See Fig. 5-22.

Stand behind gun.
Keep elbow in front.
Lean forward with
weight against gun.

Body weight must be
applied while squeezing
the trigger or the
gun will bounce off,
cutting the material or
the rivet head.

The gun is held in
one hand—the rivets
in the other.

While driving one rivet,
the experienced riveter
puts another rivet in the
next hole.

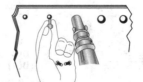

Figure 5-22 Skilled riveters develop a set procedure and work to a rhythm.

Blind bucking

In many places on an airplane structure, riveting is visually limited. A long bucking bar might have to be used and, in some cases, the bucker will not be able to see the end of the rivet. Much skill is required to do this kind of bucking in order to hold the bucking bar square with the rivet and to prevent it from coming into contact with the substructure. The driven head might have to be inspected by means of a mirror, as shown in Fig. 5-23.

Tapping code

A tapping code (Fig. 5-24) has been established to enable the rivet bucker to communicate with the mechanic driving the rivet:

1. One tap on the rivet by the rivet bucker means: start or resume driving the rivet.
2. Two taps on the rivet by the rivet bucker means that the rivet is satisfactory.
3. Three taps on the rivet by the rivet bucker means that the rivet is unsatisfactory and must be removed.

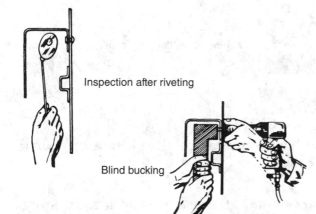

Inspection after riveting

Blind bucking

Figure 5-23 Blind bucking and inspection.

Figure 5-24 Tapping code.

CAUTION

Always tap on the rivet; do not tap on the skin or any part of the aircraft structure.

The use of modern communication methods such as two-way radios have improved the communication between the riveter and the bucker.

Hand Riveting

Hand riveting might be necessary in some cases. It is accomplished by holding a bucking bar against the rivet head, using a draw tool and a hammer to bring the sheets together, and a hand set and hammer to form the driven head (Fig. 5-25). For protruding head rivets, the bucking bar should have a cup the same size and shape as the rivet head. The hand set can be short or long, as required. Do not hammer directly on the rivet shank.

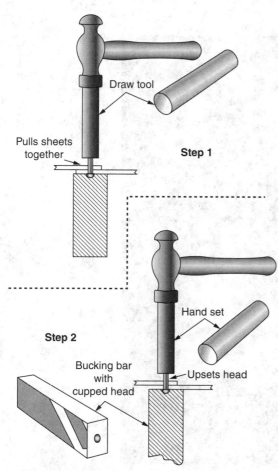

Figure 5-25   Hand riveting procedure.

Rivet Squeezers

Solid rivets can also be driven by using hand squeezers (Fig. 5-26), table riveters (Fig. 5-27), and pneumatic portable and stationary rivet squeezers (Fig. 5-28).

On some stationary squeezers, the rivets are automatically fed to the rivet sets so that the riveting operation is speeded up; on other types, the machines will punch the holes and drive the rivets as fast as the operation permits.

WARNING

Always disconnect the air hose before changing sets in a rivet squeezer.

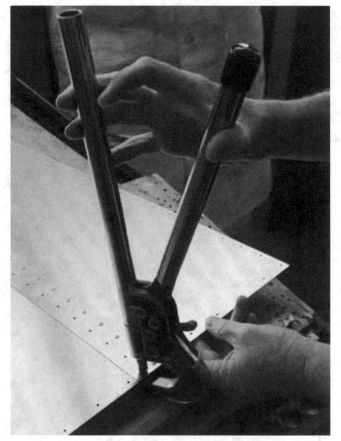

Figure 5-26   Hand rivet squeezer for small rivets.

Figure 5-27 Table riverter for dimpling and rivet installation.

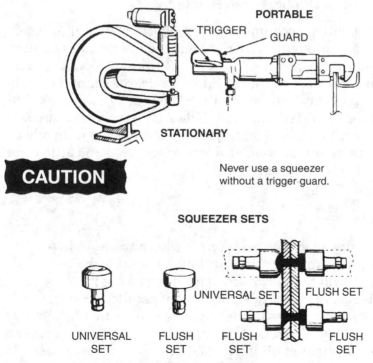

Figure 5-28 Stationary and portable rivet squeezers.

Inspection after riveting

Manufactured heads should be smooth, free of tool marks, and have no gap under the head after riveting. No cracks should be in the skin around the rivet head. The driven head should not be cocked or cracked. The height of the bucked head should be 0.5 times the rivet diameter and the width should be 1.5 times the rivet diameter. There are a few minor exceptions to these rules, but the mechanic should strive to make all rivets perfect. Figure 5-29 illustrates examples of good and bad riveting.

Rivet Removal

Solid shank rivet removal is accomplished by the procedures shown in Fig. 5-30.

NACA Method of Double Flush Riveting

A rivet installation technique known as the National Advisory Committee for Aeronautics (NACA) method has primary applications in fuel tank areas (Fig. 5-31). To make a NACA rivet installation, the shank is upset into a 82-degree countersink. In driving, the gun may be used on either the head or the shank side. The upsetting is started with light blows, then the force increased and the gun or bar moved on the shank end so as to form a head inside the countersink well. If desired, the upset head may be shaved flush after driving.

Blind Rivets

There are many places on an aircraft where access to both sides of a riveted structure or structural part is impossible, or where limited space will not permit the use of a bucking bar.

Blind rivets are rivets designed to be installed from one side of the work where access to the opposite side cannot be made to install conventional rivets. Although this was the basic reason for the development of blind rivets, they are sometimes used in applications that are not blind. This is done to save time, money, man-hours, and weight in the attachment of many nonstructural

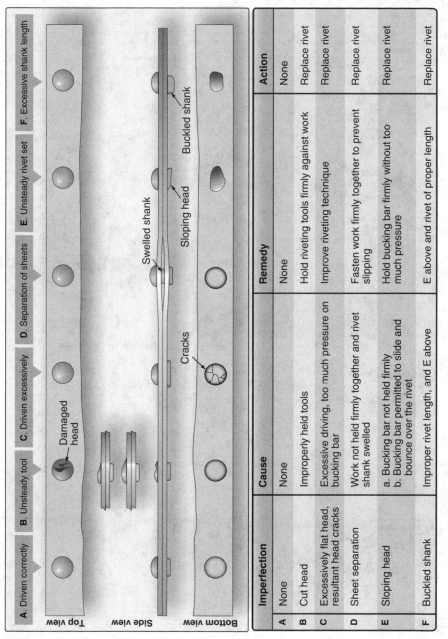

	Imperfection	Cause	Remedy	Action
A	None	None	None	None
B	Cut head	Improperly held tools	Hold riveting tools firmly against work	Replace rivet
C	Excessively flat head, resultant head cracks	Excessive driving, too much pressure on bucking bar	Improve riveting technique	Replace rivet
D	Sheet separation	Work not held firmly together and rivet shank swelled	Fasten work firmly together to prevent slipping	Replace rivet
E	Sloping head	a. Bucking bar not held firmly b. Bucking bar permitted to slide and bounce over the rivet	Hold bucking bar firmly without too much pressure	Replace rivet
F	Buckled shank	Improper rivet length, and E above	E above and rivet of proper length	Replace rivet

Figure 5-29 Typical rivet defects.

Rivet Removal

Remove rivets by drilling off the head and punching out the shank as illustrated.
1. File a flat area on the manufactured head of non-flush rivets.
2. Place a block of wood or a bucking bar under both flush and nonflush rivets when center punching the manufactured head.
3. Use a drill that is ⅟₃₂ (0.0312) inch smaller than the rivet shank to drill through the head of the rivet. Ensure the drilling operation does not damage the skin or cut the sides of the rivet hole.
4. Insert a drift punch into the hole drilled in the rivet and tilt the punch to break off the rivet head.
5. Using a drift punch and hammer, drive out the rivet shank. Support the opposite side of the structure to prevent structural damage.

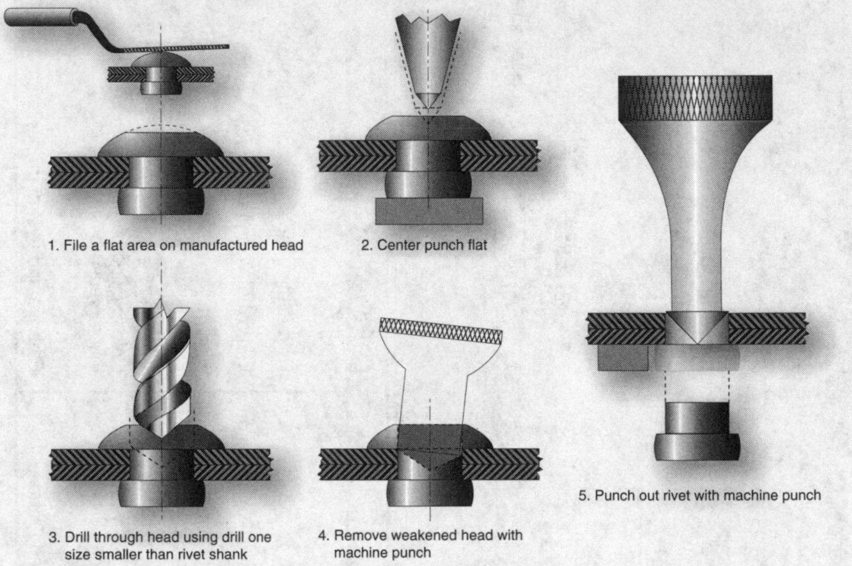

1. File a flat area on manufactured head

2. Center punch flat

3. Drill through head using drill one size smaller than rivet shank

4. Remove weakened head with machine punch

5. Punch out rivet with machine punch

Figure 5-30   Solid shank rivet removal procedures.

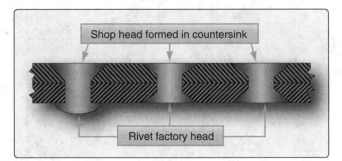

Figure 5-31   NACA riveting method.

parts, such as aircraft interior furnishings, flooring, deicing boots, and the like, where the full strength of solid-shank rivets is not necessary. These rivets are produced by several manufacturers and have unique characteristics that require special installation tools, special installation procedures, and special removal procedures.

Basically, nearly all blind rivets depend upon the principle of drawing a stem or mandrel through a sleeve to accomplish the forming of the bucked (upset) head.

Although many variations of blind rivets exist, depending on the manufacturer, there are essentially three types:

- Hollow, pull-through rivets (Fig. 5-32), used mainly for non-structural applications.

- Self-plugging, friction-lock rivets (Fig. 5-26), whereby the stem is retained in the rivet by friction. Although strength of these rivets approaches that of conventional solid-shank rivets, there is no positive mechanical lock to retain the stem.

- Mechanical locked-stem self-plugging rivets (Fig. 5-33), whereby a locking collar mechanically retains the stem in the rivet.

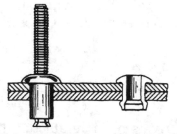

Before installation   After installation

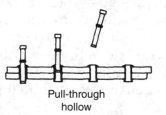

Pull-through
hollow

Figure 5-32 Pull-through rivets (hollow).

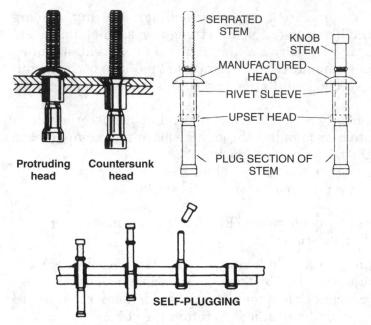

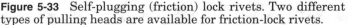

Figure 5-33 Self-plugging (friction) lock rivets. Two different types of pulling heads are available for friction-lock rivets.

This positive lock resists vibration that could cause the friction-lock rivets to loosen and possibly fall out. Self-plugging mechanical-lock rivets display all the strength characteristics of solid-shank rivets; in almost all cases, they can be substituted rivet for rivet.

Mechanical locked-stem self-plugging rivets

Mechanical locked-stem self-plugging rivets are manufactured by Olympic, Huck, and Cherry Fasteners. The bulbed CherryMax® (Fig. 5-34) is used as an example of a typical blind rivet that is virtually interchangeable, structurally, with solid rivets.

The installation of all mechanical locked-stem self-plugging rivets requires hand or pneumatic pull guns with appropriate pulling heads. Many types are available from the rivet manufacturers; examples of hand and pneumatic-operated pull guns are shown in Fig. 5-35.

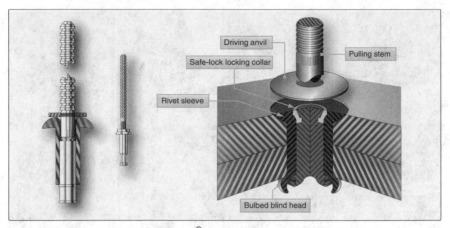

Figure 5-34 The bulbed CherryMax® rivet includes a locking collar to firmly retain the portion of the stem in the rivet sleeve.

Figure 5-35 CherryMax® rivet puller.

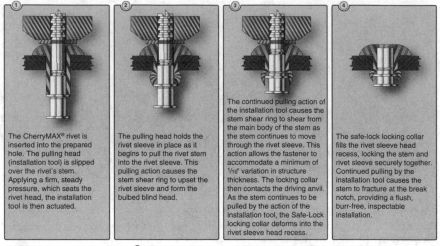

① The CherryMAX® rivet is inserted into the prepared hole. The pulling head (installation tool) is slipped over the rivet's stem. Applying a firm, steady pressure, which seats the rivet head, the installation tool is then actuated.

② The pulling head holds the rivet sleeve in place as it begins to pull the rivet stem into the rivet sleeve. This pulling action causes the stem shear ring to upset the rivet sleeve and form the bulbed blind head.

③ The continued pulling action of the installation tool causes the stem shear ring to shear from the main body of the stem as the stem continues to move through the rivet sleeve. This action allows the fastener to accommodate a minimum of 1/16″ variation in structure thickness. The locking collar then contacts the driving anvil. As the stem continues to be pulled by the action of the installation tool, the Safe-Lock locking collar deforms into the rivet sleeve head recess.

④ The safe-lock locking collar fills the rivet sleeve head recess, locking the stem and rivet sleeve securely together. Continued pulling by the installation tool causes the stem to fracture at the break notch, providing a flush, burr-free, inspectable installation.

Figure 5-36   CherryMax® blind rivet installation.

The sequence of events in forming the bulbed CherryMax® rivet is shown in Fig. 5-36. Figure 5-37 illustrates the numbering system for bulbed CherryMax® rivets.

Hole preparation.   The bulbed CherryMax® rivets are designed to function within a specified hole size range and countersink dimensions as listed in Fig. 5-38.

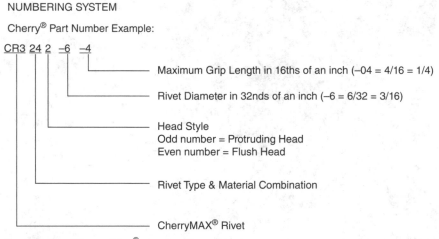

NUMBERING SYSTEM

Cherry® Part Number Example:

CR3 24 2 –6 –4

Maximum Grip Length in 16ths of an inch (–04 = 4/16 = 1/4)

Rivet Diameter in 32nds of an inch (–6 = 6/32 = 3/16)

Head Style
Odd number = Protruding Head
Even number = Flush Head

Rivet Type & Material Combination

CherryMAX® Rivet

Figure 5-37   CherryMax® numbering system.

Nominal CherryMAX

Rivet Diameter	Drill Size	Hole Size	
		Min.	Max.
-4 (1/8")	#30	.129	.132
-5 (5/32")	#20	.160	.164
-6 (3/16")	#10	.192	.196
-8 (1/4")	F	.256	.261

Oversize CherryMAX

Rivet Diameter	Drill Size	Hole Size	
		Min.	Max.
-4 (1/8")	#27	.143	.146
-5 (5/32")	#16	.176	.180
-6 (3/16")	#5	.205	.209
-8 (1/4")	1	.271	.275

Rivet Diameter	MS 20426 100° Head		NAS1097 100° Head		Unisink 100° Head		120° Head	
	C MIN.	C MAX.	C MIN.	C MAX.	C MIN.	C MAX.	C MIN.	C MAX.
-4 (1/8")	.222	.228	.189	.195	.167	.173	.269	.275
-5 (5/32")	.283	.289	.240	.246	.210	.216	.311	.317
-6 (3/16")	.350	.356	.296	.302	.252	.258	.347	.353
-8 (1/4")	.473	.479	.389	.395	—	—	—	—

Figure 5-38 Recommended drill-sized, hole-sized, and countersunk diameter limits.

Grip length.   *Grip length* refers to the maximum total sheet thickness to be riveted, and is measured in 16ths of an inch. This is identified by the second dash number. All CherryMax® rivets have their grip length (maximum grip) marked on the rivet head, and have a total grip range of ⅟₁₆ of an inch. (Example: A -4 grip rivet has a grip range of 0.188 to 0.250 inch; Fig. 5-39.) To determine the proper grip rivet to use, measure the material thickness with a Cherry® selector gauge, as shown in Fig. 5-40. Always read to the next higher number.

Further data on bulbed CherryMax® rivets, including materials available, is included in Chap. 13, Standard Parts.

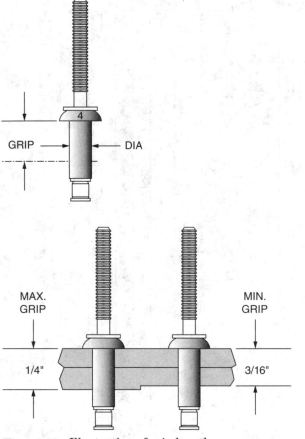

Figure 5-39   Illustration of grip length.

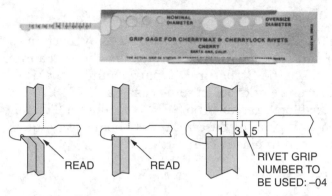

READ   READ   RIVET GRIP
                            NUMBER TO
                            BE USED: –04

Figure 5-40   Determining the proper grip using a grip gage.

Complete installation manuals and pulling tool catalogs are available from the rivet manufacturers.

Removal of mechanically locked blind rivets

Mechanically locked blind rivets are a challenge to remove because they are made from strong, hard metals. Lack of access poses yet another problem for the aviation technician. Designed for and used in difficult to reach locations means there is often no access to the blind side of the rivet or any way to provide support for the sheet metal surrounding the rivet's location when the aviation technician attempts removal. The stem is mechanically locked by a small lock ring that needs to be removed first. Use a small center drill to provide a guide for a larger drill on top of the rivet stem and drill away the upper portion of the stem to destroy the lock. Try to remove the lock ring or use a prick punch or center punch to drive the stem down a little and remove the lock ring. After the lock ring is removed, the stem can be driven out with a drive punch. After the stem is removed, the rivet can be drilled out in the same way as a solid rivet. If possible, support the back side of the rivet with a backup block to prevent damage to the aircraft skin.

Sheet-Metal Repair

Metal aircraft get damaged and need to be repaired by mechanics and returned to service and airworthy status. The mechanic should always consult the Structural Repair Manual (SRM) of the airplane that he/she works on. In some cases, the AC43.13-1B could be used to develop a repair if no SRM is available. The following discussion is a general description of repair techniques and should not be used in place of the SRM.

Damage removal

To prepare a damaged area for repair:

1. Remove all distorted skin and structure in damaged area.
2. Remove damaged material so that the edges of the completed repair match existing structure and aircraft lines.
3. Round all square corners.
4. Smooth out any abrasions and/or dents.
5. Remove and incorporate into the new repair any previous repairs joining the area of the new repair.

Repair material thickness

The repair material must duplicate the strength of the original structure. If an alloy weaker than the original material has to be used, a heavier gauge must be used to give equivalent cross-sectional strength. A lighter gauge material should not be used even when using a stronger alloy. At times, the original material is not available and the material needs to be substituted by another alloy. Figure 5-41 shows a material substitution chart.

Rivet selection

Normally, the rivet size and material should be the same as the original rivets in the part being repaired. If a rivet hole has been enlarged or deformed, the next larger size rivet must be used after reworking the hole. When this is done, the proper edge

Shape	Initial Material	Replacement Material
Sheet 0.016 to 0.125	Clad 2024–T42 Ⓕ	Clad 2024–T3 / 2024–T3 / Clad 7075–T6 Ⓐ / 7075–T6 Ⓐ
	Clad 2024–T3	2024–T3 / Clad 7075–T6 Ⓐ / 7075–T6 Ⓐ
	Clad 7075–T6	7075–T6
Formed or extruded section	2024–T42 Ⓕ	7075–T6 Ⓐ Ⓑ

Sheet Material To Be Replaced	Material Replacement Factor									
	7075–T6	Clad 7075–T6	2024–T3		Clad 2024–T3		Ⓕ 2024–T4 2024–T42		Ⓕ Clad 2024–T4 Clad 2024–T42	
	Ⓒ	Ⓒ Ⓗ	Ⓓ	Ⓔ	Ⓓ	Ⓔ	Ⓓ	Ⓔ	Ⓓ	Ⓔ
7075–T6	1.00	1.10	1.20	1.78	1.30	1.83	1.20	1.78	1.24	1.84
Clad 7075–T6	1.00	1.00	1.13	1.70	1.22	1.76	1.13	1.71	1.16	1.76
2024–T3	1.00 Ⓐ	1.00 Ⓐ	1.00	1.00	1.09	1.10	1.00	1.10	1.03	1.14
Clad 2024–T3	1.00 Ⓐ	1.00 Ⓐ	1.00	1.00	1.00	1.00	1.00	1.00	1.03	1.00
2024–T42	1.00 Ⓐ	1.00 Ⓐ	1.00	1.00	1.00	1.00	1.00	1.00	1.00	1.14
Clad 2024–T42	1.00 Ⓐ	1.00 Ⓐ	1.00	1.00	1.00	1.00	1.00	1.00	1.00	1.00
7178–T6	1.28	1.28	1.50	1.90	1.63	2.00	1.86	1.90	1.96	1.98
Clad 7178–T6	1.08	1.18	1.41	1.75	1.52	1.83	1.75	1.75	1.81	1.81
5052–H34 Ⓖ Ⓗ	1.00 Ⓐ	1.00 Ⓐ	1.00	1.00	1.00	1.00	1.00	1.00	1.00	1.00

Notes:
- All dimensions are in inches, unless otherwise specified.

- It is possible that more protection from corrosion is necessary when bare mineral is used to replace clad material.

- It is possible for the material replacement factor to be a lower value for a specific location on the airplane. To get that value, contact Boeing for a case-by-case analysis.

- Example:
  To refer 0.040 thick 7075–T6 with clad 7075–T6, multiply the gauge by the material replacement factor to get the replacement gauge
  0.040 × 1.10 = 0.045.

Ⓐ Cannot be used as replacement for the initial material in areas that are pressured.

Ⓑ Cannot be used in the wing interspar structure at the wing center section structure.

Ⓒ Use the next thicker standard gauge when using a formed section as a replacement for an extrusion.

Ⓓ For all gauges of flat sheet and formed sections.

Ⓔ For flat sheet < 0.071 thick.

Ⓕ For flat sheet ≥ 0.071 thick and for formed sections.

Ⓖ 2024–T4 and 2024–T42 are equivalent.

Ⓗ A compound to give protection from corrosion must be applied to bare material that is used to replace 5052–H34.

Figure 5-41 Material substitution chart.

distance for the larger rivet must be maintained. Where access to the inside of the structure is impossible and blind rivets must be used in making the repair, always consult the applicable aircraft maintenance manual for the recommended type, size, spacing, and number of rivets needed to replace either the original installed rivets or those that are required for the type of repair being performed. As a general rule, the rivet diameter is three times the thickness of the thicker sheet. For example, if the repair material is .040 inch thick, the rivet must be $3 \times 0.032 = 0.120$ inch. Select a ⅛ (0.125) rivet diameter.

Rivet spacing and edge distance

The rivet pattern for a repair must conform to instructions in the applicable aircraft manual. The existing rivet pattern on the aircraft is used whenever possible. The typical rivet spacing is 4 to 6D. Figures 5-8 and 5-11 contain information about minimum acceptable rivet spacing and edge distance. Figure 5-42 shows different ways to layout a rivet pattern.

Repair approval

FAA Form 337, Major Repair and Alteration, must be completed for repairs to the following parts of an airframe and repairs of the following types involving the strengthening, reinforcing, splicing, and manufacturing of primary structural members or their replacement, when replacement is by fabrication, such as riveting or welding.

- Box beams
- Monocoque or semimonocoque wings or control surfaces
- Wing stringers or chord members
- Spars
- Spar flanges
- Members of truss-type beams
- Thin sheet webs of beams

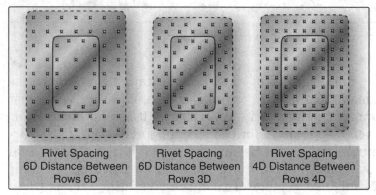

Figure 5-42   Rivet spacing layout.

- Keel and chine members of boat hulls or floats
- Corrugated sheet compression members that act as flange material of wings or tail surfaces
- Wing main ribs and compression members
- Wing or tail surface brace struts, fuselage longerons
- Members of the side truss, horizontal truss, or bulkheads
- Main seat support braces and brackets
- Landing gear brace struts
- Repairs involving the substitution of material
- Repair of damaged areas in metal or plywood stressed covering exceeding six inches in any direction
- Repair of portions of skin sheets by making additional seams
- Splicing of thin sheets
- Repair of three or more adjacent wing or control surface ribs or the leading edge of wings and controlsurfaces between such adjacent ribs

For major repairs made in accordance with a manual or specifications acceptable to the Administrator, a certificated repair station may use the customer's work order upon which the repair is recorded in place of the FAA Form 337.

Typical sheet-metal repairs

The following section will discuss several standard type of repairs of aircraft sheet-metal structures.

Patches

Skin patches may be classified as two types: lap or scab patch and flush patch.

Lap or scab patch. The lap or scab type of patch is an external patch where the edges of the patch and the skin overlap each other. The overlapping portion of the patch is riveted to the skin. Lap patches may be used in most areas where aerodynamic smoothness is not important. Figure 5-43 shows a typical patch for a crack and/or for a hole.

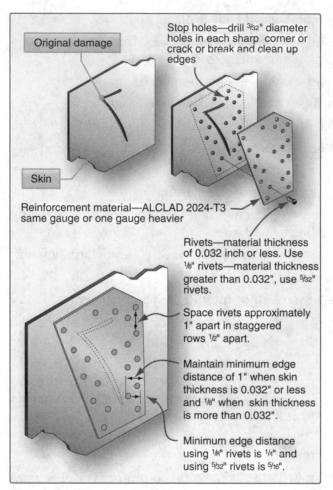

Stop holes—drill ³⁄₃₂" diameter holes in each sharp corner or crack or break and clean up edges

Original damage

Skin

Reinforcement material—ALCLAD 2024-T3 same gauge or one gauge heavier

Rivets—material thickness of 0.032 inch or less. Use ⅛" rivets—material thickness greater than 0.032", use ⁵⁄₃₂" rivets.

Space rivets approximately 1" apart in staggered rows ½" apart.

Maintain minimum edge distance of 1" when skin thickness is 0.032" or less and ⅛" when skin thickness is more than 0.032".

Minimum edge distance using ⅛" rivets is ¼" and using ⁵⁄₃₂" rivets is ⁵⁄₁₆".

Figure 5-43   Lap or scab patch.

Flush Patch

A flush patch is a filler patch that is flush to the skin when applied, it is supported by and riveted to a reinforcement plate which is, in turn, riveted to the inside of the skin. Figure 5-44 shows a typical flush patch repair of a leading edge and Fig. 5-45 shows a flush repair of a nonpressurized structure. The doubler is inserted

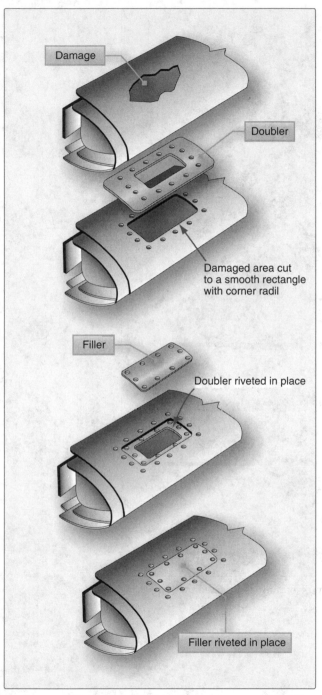

Figure 5-44 Flush repair of leading edge.

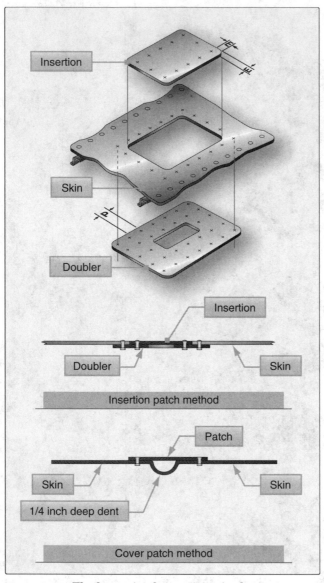

Figure 5-45 Flush repair of nonpressurized structure.

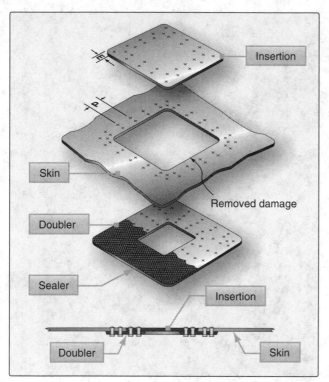

Figure 5-46 Pressurized skin repair.

through the opening and rotated until it slides in place under the skin. The filler must be of the same gauge and material as the original skin. The doubler should be of material one gauge heavier than the skin.

Mechanics should always consult the SRM of the aircraft they are working on but the following figures show some typical repairs to illustrate what those repairs look like. Figure 5-46 shows the details of a typical repair of a pressurized structure, Fig. 5-47 shows the details of a bulk head repair, and Fig. 5-48 shows the details of a wing-rib and lightening hole repair.

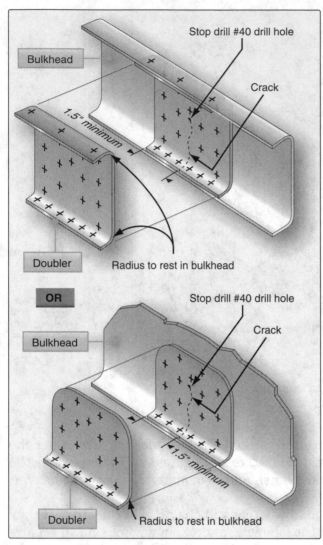

Figure 5-47 Bulk-head repair.

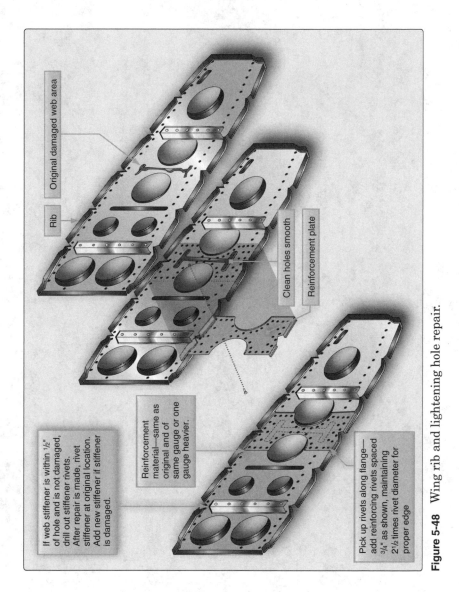

Original damaged web area

Rib

Clean holes smooth

Reinforcement plate

If web stiffener is within 1/2" of hole and is not damaged, drill out stiffener rivets. After repair is made, rivet stiffener at original location. Add new stiffener if stiffener is damaged.

Reinforcement material—same as original and of same gauge or one gauge heavier.

Pick up rivets along flange—add reinforcing rivets spaced 3/4" as shown, maintaining 2½ times rivet diameter for proper edge

Figure 5-48 Wing rib and lightening hole repair.

6

Bolts and Threaded Fasteners

Various types of fastening devices allow quick dismantling or replacement of aircraft parts that must be taken apart and put back together at frequent intervals. Bolts and screws are two types of fastening devices that give the required security of attachment and rigidity. Generally, bolts are used where great strength is required, and screws are used where strength is not the deciding factor.

The threaded end of a bolt usually has a nut screwed onto it to complete the assembly. The threaded end of a screw might fit into a female receptacle, or it might fit directly into the material being secured. A bolt has a fairly short threaded section and a comparatively long grip length or unthreaded portion, whereas a screw has a longer threaded section and might have no clearly defined grip length. A bolt assembly is generally tightened by turning the nut on the bolt; the head of the bolt might not be designed for turning. A screw is always tightened by turning its head.

The modern high-performance jet aircraft, however, uses very few "standard" hex head bolts and nuts in its assembly. Also, the "standard" slotted and Phillips head screws are in the minority. Some of these advanced fasteners are described later in this chapter.

In many cases, a bolt might be indistinguishable from a screw, thus the term *threaded fastener*. Also, many threaded fasteners, such as the Hi-Lok® and Hi-Lok®/Hi-Tigue® fasteners, are essentially permanent installations, like a rivet.

Aircraft threaded fasteners are fabricated from alloy steel, corrosion-resistant (stainless) steel, aluminum alloys, and titanium. Most bolts used in aircraft are either alloy steel, cadmium plated, general-purpose AN bolts, NAS close-tolerance, or MS bolts. Aluminum bolts are seldom used in the primary structure. In certain cases, aircraft manufacturers make threaded fasteners of different dimensions or greater strength than the standard types. Such threaded fasteners are made for a particular application, and it is of extreme importance to use similar fasteners in replacement.

Aircraft Bolts

Most, but not all, aircraft bolts are designed and fabricated according to government standards with the following specifications:

- AN, Air Force/Navy
- NAS, National Aerospace Standards
- MS, Military Standards

See Chap. 14, Standard Parts, for more information concerning government standards.

General-purpose bolts

The hex-head aircraft bolt (AN-3 through AN-20) is an all-purpose structural bolt used for general applications involving tension or shear loads where a light-drive fit is permissible (0.006-inch clearance for a ⅝-inch hole, and other sizes in proportion). They are fabricated from SAE 2330 nickel steel and are cadmium plated.

Alloy steel bolts smaller than No. 10-32 (³⁄₁₆-inch diameter, AN-3) and aluminum alloy bolts smaller than ¼-inch diameter are not used in primary structures. Aluminum alloy bolts and nuts are not used where they will be repeatedly removed for purposes of maintenance and inspection.

The AN73-AN81 (MS20073-MS20074) drilled-head bolt is similar to the standard hex-bolt, but has a deeper head that is drilled to receive wire for safetying. The AN3-AN20 and the AN73-AN81 series bolts are interchangeable, for all practical purposes, from the standpoint of tension and shear strengths (see Chap. 14, Standard Parts).

Close-Tolerance Bolts

This type of bolt is machined more accurately than the general-purpose bolt. Close-tolerance bolts can be hex-headed (AN-173 through AN-186) or have a 100-degree countersunk head (NAS-80 through NAS-86). They are used in applications where a tight drive fit is required (the bolt will move into position only when struck with a 12- to 14-ounce hammer).

Classification of Threads

Aircraft bolts, screws, and nuts are threaded in either the NC (American National Coarse) thread series, the NF (American National Fine) thread series, the UNC (American Standard Unified Coarse) thread series, or the UNF (American Standard Unified Fine) thread series. Threads are designated by the number of times the incline (threads) rotates around a 1-inch length of a given diameter bolt or screw. For example, a 4-28 thread indicates that a ¼-inch-diameter bolt has 28 threads in 1 inch of its threaded length.

Threads are also designated by the class of fit. The class of a thread indicates the tolerance allowed in manufacturing. Class 1 is a loose fit, class 2 is a free fit, class 3 is a medium fit, and class 4 is a close fit. **Aircraft bolts are almost always manufactured in the class 3, medium fit.** A class-4 fit requires a wrench to turn the nut onto a bolt, whereas a class-1 fit can easily be turned with the fingers. Generally, aircraft screws are manufactured with a class-2 thread fit for ease of assembly. The general-purpose aircraft bolt, AN-3 through AN-20 has UNF-3 threads (American Standard Unified Fine, class 3, medium fit).

Bolts and nuts are also produced with right-hand and left-hand threads. A right-hand thread tightens when turned clockwise; a left-hand thread tightens when turned counterclockwise. Except in special cases, all aircraft bolts and nuts have right-hand threads.

Identification and coding

Threaded fasteners are manufactured in many shapes and varieties. A clear-cut method of classification is difficult. Threaded fasteners can be identified by the shape of the head, method of securing, material used in fabrication, or the expected usage. Figure 6-1 shows the basic head styles and wrenching recesses.

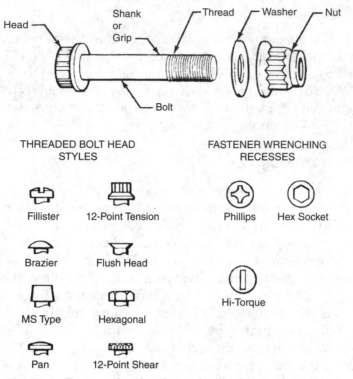

Figure 6-1   Fastener head styles and wrenching recesses.

AN-type aircraft bolts can be identified by the code markings on the boltheads. The markings generally denote the bolt manufacturer, composition of the bolt, and whether the bolt is a standard AN-type or a special-purpose bolt. AN standard steel bolts are marked with either a raised dash or asterisk (Fig. 6-2), corrosion-resistant steel is indicated by a single raised dash, and AN aluminum alloy bolts are marked with two raised dashes. Additional information, such as bolt diameter, bolt length, and grip length can be obtained from the bolt part number. See Chap. 13, Standard Parts.

MS and NAS style bolts often show the partnumber on the head and are readily identified.

Aircraft Nuts

Aircraft nuts are manufactured in a variety of shapes and sizes, made of alloy steel, stainless steel, aluminum alloy, brass, or titanium.

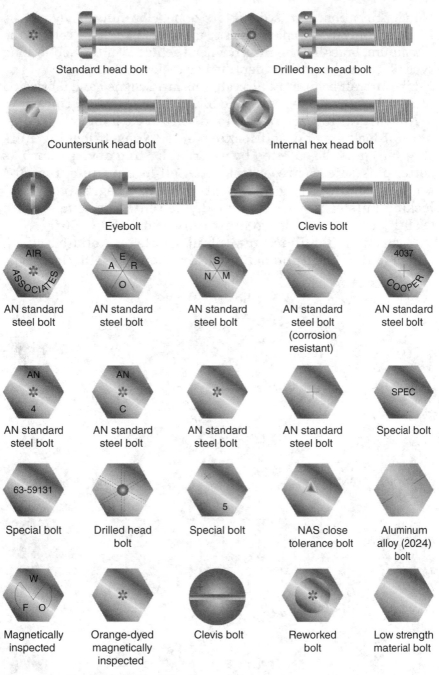

Figure 6-2 Aircraft bolt identification.

No identification marks or letters appear on nuts. They can be identified only by the characteristic metallic luster or by color of the aluminum, brass, or the insert, when the nut is of the self-locking type. They can be further identified by their construction.

Like aircraft bolts, most aircraft nuts are designed and fabricated in accordance with AN, NAS, and MS standards and specifications.

Aircraft nuts can be divided into two general groups: nonself-locking and self-locking nuts. Nonself-locking nuts (Fig. 6-3) must be safetied by external locking devices, such as cotter pins, safety wire, or locknuts. Self-locking nuts (Figs. 6-4 and 6-5) contain the locking feature as an integral part. Self-locking nuts can be further subdivided into low temperature (250°F or less) and high temperature (more than 250°F).

Most of the familiar nuts (plain, castle, castellated shear, plain hex, light hex, and plain check) are the nonselflocking type (Fig. 6-3).

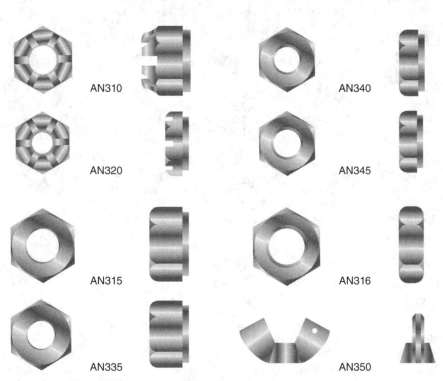

Figure 6-3 Nonself-locking, castellated, and plain nuts.

| FLEXLOC® Hex Self-Locking Regular Height | Hexagon Self-Locking Castellated Nut | 12 Point Self-Locking Tension Nut | 12 Point Self-Locking Shear Nut | 12 Spline Self-Locking Tension Nut | 12 Spline Self-Locking Shear Nut |

Figure 6-4 High-temperature (more than 250°F) self-locking nuts.

The castle nut, AN-310, is used with drilled-shank AN hex-head bolts, clevis bolts, eyebolts, drilled head bolts, or studs. It is fairly rugged and can withstand large-tension loads. Slots (castellations) in the nut are designed to accommodate a cotter pin or lock wire for safety. The AN-310 castellated, cadmium-plated steel nut is by far the most commonly used airframe nut. See Chap. 13, Standard Parts.

The castellated shear nut, AN-320, is designed for use with devices (such as drilled clevis bolts and threaded taper pins) that are normally subjected to shearing stress only. Like the castle nut, it is castellated for safetying. Note, however, that the nut is not as deep or as strong as the castle nut; also notice that the castellations are not as deep as those in the castle nut.

Self-locking nuts to 250°F

The elastic stop nut is essentially a standard hex nut that incorporates a fiber or nylon insert (Fig. 6-5). The inside diameter of the red insert is deliberately smaller than the major diameter of the matching bolt. The nut spins freely on the bolt until the bolt threads enter the locking insert, where they impress, but do not cut, mating threads in the insert. This compression forces a metal-to-metal contact between the top flanks of the nut threads and the bottom flanks of the bolt threads. This friction

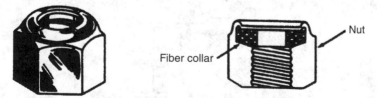

Fiber collar

Nut

Figure 6-5 Low-temperature (250°F or less) self-locking nut (elastic stop nut, AN365, MS20365).

hold plus the compression hold of the insert essentially "locks" the nut anywhere on the bolt.

After the nut has been tightened, the rounded or chamfered end of bolts, studs, or screws should extend at least the full round or chamfer through the nut. Flat-end bolts, studs, or screws should extend at least ½₂ inch through the nut. When fiber-type self-locking nuts are reused, the fiber should be carefully checked to be sure that it has not lost its locking friction or become brittle. Locknuts should not be reused if they can be run up to a finger-tight position. Bolts ⁵⁄₁₆ inch diameter and larger, with cotter pin holes, can be used with self-locking nuts, but only if they are free from burrs around the holes. Bolts with damaged threads and rough ends are not acceptable.

Self-locking nuts should not be used at joints that subject either the nut or the bolt to rotation. They can be used with antifriction bearings and control pulleys, provided that the inner face of the bearing is clamped to the supporting structure by the nut and bolt.

High-temperature self-locking nuts

All-metal locknuts are constructed with either the threads in the locking insert out-of-phase with the load-carrying section (Fig. 6-6) or with a saw-cut insert with a pinched-in thread in the locking section. The locking action of the all-metal nut depends upon the

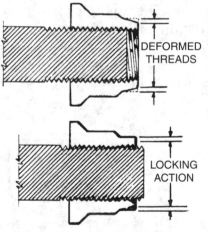

Figure 6-6 The Boot's self-locking, all-metal nut.

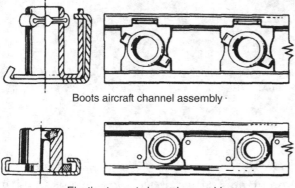

Boots aircraft channel assembly ·

Elastic stop nut channel assembly

Figure 6-7 Self-locking nut bases.

resiliency of the metal when the locking section and load-carrying section are engaged by screw threads.

Miscellaneous nut types

Self-locking nut bases are made in a number of forms and materials for riveting and welding to aircraft structure or parts (Fig. 6-7). Certain applications require the installation of self-locking nuts in channels, an arrangement that permits the attachment of many nuts with only a few rivets. These channels are track-like bases with regularly spaced nuts that are either removable or nonremovable. The removable type carries a floating nut that can be snapped in or out of the channel, thus making possible the easy removal of damaged nuts. Clinch and spline nuts, which depend on friction for their anchorage, are not acceptable for use in aircraft structures.

Various types of anchor nuts (Fig. 6-8) are available for riveting to the structure for application as removable panels.

Figure 6-8 Examples of anchor nuts.

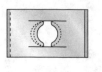

FLAT TYPE

ANCHOR TYPE

"U" TYPE

Figure 6-9   Sheet spring nuts are used with self-tapping screws in non-structural locations.

Sheet spring nuts, sometimes called *speed nuts*, are used with standard and sheet-metal self-tapping screws in nonstructural locations. They find various uses in supporting line clamps, conduit clamps, electrical equipment access doors, etc., and are available in several types. Speed nuts are made from spring steel and are arched prior to tightening. This arched spring lock prevents the screw from working loose. These nuts should be used only where originally used in fabrication of the aircraft (Fig. 6-9).

Aircraft Washers

Aircraft washers used in airframe repair are plain, lock, or special washers.

Plain washers

The plain washer, AN-960 (Fig. 6-10), is used under hex nuts. It provides a smooth bearing surface and acts as a shim in obtaining correct grip length for a bolt and nut assembly. It is used to adjust the position of castellated nuts with respect to drilled cotter pin holes in bolts. Plain washers should be used under lock washers to prevent damage to the surface material.

Lock washers

Lock washers (AN-935 and AN-936) can be used with machine screws or bolts whenever the self-locking or castellated nut is not applicable. They are not to be used as fastenings to primary or secondary structures, or where subject to frequent removal or corrosive conditions.

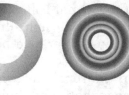

Plain AN 960   Ball seat & socket   Taper pin
                 AC9950 & AC955   AN975

Special washers

Plain AN 935   Star lock washers

Figure 6-10 Plain and lock washers.

Installation of Nuts and Bolts

Boltholes must be normal to the surface involved to provide full bearing surface for the bolthead and nut and must not be oversized or elongated. A bolt in such a hole will carry none of its shear load until parts have yielded or deformed enough to allow the bearing surface of the oversized hole to contact the bolt.

In cases of oversized or elongated holes in crucial members, obtain advice from the aircraft or engine manufacturer before drilling or reaming the hole to take the next larger bolt. Usually, such factors as edge distance, clearance, or load factor must be considered. Oversized or elongated holes in noncrucial members can usually be drilled or reamed to the next larger size.

Many boltholes, particularly those in primary connecting elements, have close tolerances. Generally, it is permissible to use the first lettered drill size larger than the normal bolt diameter, except where the AN hexagon bolts are used in light-drive fit (reamed) applications and where NAS close-tolerance bolts or AN clevis bolts are used.

Light-drive fits for bolts (specified on the repair drawings as .0015-inch maximum clearance between bolt and hole) are required in places where bolts are used in repair, or where they are placed in the original structure.

The fit of holes and bolts is defined in terms of the friction between the bolt and hole when sliding the bolt into place. A tight-drive fit, for example, is one in which a sharp blow of a 12- or 14-ounce hammer is required to move the bolt. A bolt that requires a hard blow and sounds tight is considered to fit too tightly. A light-drive fit is one in which a bolt will move when a hammer handle is held against its head and pressed by the weight of the body.

Examine the markings on the bolthead to determine that each bolt is of the correct material. It is of extreme importance to use similar bolts in replacement. In every case, refer to the applicable maintenance instruction manual and the illustrated parts breakdown.

Be sure that washers are used under the heads of bolts and nuts, unless their omission is specified. A washer guards against mechanical damage to the material being bolted and prevents corrosion of the structural members.

Be certain that the bolt grip length is correct. The grip length is the length of the unthreaded portion of the bolt shank (Fig. 6-11). Generally speaking, the grip length should equal the thickness of the materials being bolted together. However, bolts of slightly greater grip length can be used if washers are placed under the nut or the bolthead. In the case of plate nuts, add shims under the plate.

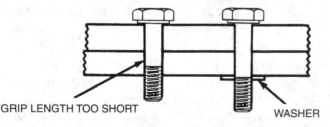

GRIP LENGTH TOO SHORT   WASHER

Figure 6-11   Bolt installation.

A nut is not run to the bottom of the threads on the bolt. A nut so installed cannot be pulled tight on the structure and probably will be twisted off while being tightened. A washer will keep the nut in the proper position on the bolt.

In the case of self-locking stop nuts, if from one to three threads of the bolt extend through the nut, it is considered to be satisfactory (Fig. 6-12).

Palnuts (AN-356) should be tightened securely, but not excessively. Finger-tight plus one to two turns is good practice, two turns being more generally used.

Torque tables

The standard torque table (Fig. 6-13a) should be used as a guide in tightening nuts, studs, bolts, and screws whenever specific torque values are not caged out in maintenance procedures.

Calculate the correct torque value using the information in Fig. 6-13b if an extension needs to be used with the torque wrench. The extension could alter the actual torque value.

Cotter pin hole line-up

When tightening castellated nuts on bolts, the cotter pin holes may not line up with the slots in the nuts for the range of recommended values. Except in cases of highly stressed engine parts, the nut may not be over torqued. Remove hardware and realign

BOLT OR SCREW SIZE	MINIMUM BOLT PROTRUSION THROUGH NUT ('A' DIMENSION)
3/16	0.062
1/4	0.072
5/16 and 3/8	0.083
7/16 and 1/2	0.100
9/16 and 5/8	0.110
3/4	0.125
7/8	0.140
1 to 1-1/2	0.165

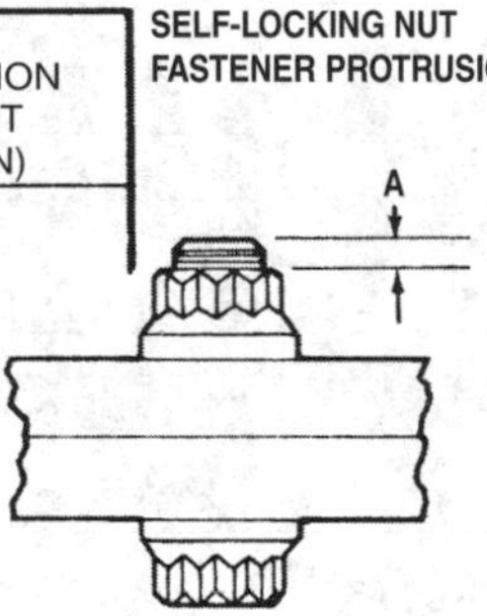

SELF-LOCKING NUT
FASTENER PROTRUSION

A

Figure 6-12   Minimum bolt protrusion through the nut. Note: Do not use self-locking nuts on bolts drilled for cotter pins.

Bolt, Stud, or Screw Size		Torque Values in Inch-Pounds for Tightening Nuts			
		On standard bolts, studs and screws having a tensile strength of 125,000 to 140,000 psi		On bolts, studs, and screws having a tensile strength of 140,000 to 160,000 psi	On high-strength bolts, studs, and screws having a tensile strength of 160,000 psi and over
		Shear type nuts (AN320, AN364 or equivalent)	Tension type nuts and threaded machine parts (AN-310, AN365, or equivalent)	Any nut, except shear type	Any nut, except shear type
8–32	8–36	7–9	12–15	14–17	15–18
10–24	10–32	12–15	20–25	23–30	25–35
¼–20		25–30	40–50	45–49	50–68
	¼–28	30–40	50–70	60–80	70–90
⁵⁄₁₆–18		48–55	80–90	85–117	90–144
	⁵⁄₁₆–24	60–85	100–140	120–172	140–203
³⁄₈–16		95–110	160–185	173–217	185–248
	³⁄₈–24	95–110	160–190	175–271	190–351
⁷⁄₁₆–14		140–155	235–255	245–342	255–428
	⁷⁄₁₆–20	270–300	450–500	475–628	500–756
½–13		240–290	400–480	440–636	480–792
	½–20	290–410	480–690	585–840	690–990
⁹⁄₁₆–12		300–420	500–700	600–845	700–990
	⁹⁄₁₆–18	480–600	800–1000	900–1,220	1,000–1,440
⁵⁄₈–11		420–540	700–900	800–1,125	900–1,350
	⁵⁄₈–18	660–780	1,100–1,300	1,200–1,730	1,300–2,160
¾–10		700–950	1,150–1,600	1,380–1,925	1,600–2,250
	¾–16	1,300–1,500	2,300–2,500	2,400–3,500	2,500–4,500
⁷⁄₈–9		1,300–1,800	2,200–3,000	2,600–3,570	3,000–4,140
	⁷⁄₈–14	1,500–1,800	2,500–3,000	2,750–4,650	3,000–6,300
1"–8		2,200–3,000	3,700–5,000	4,350–5,920	5,000–6,840
	1"–14	2,200–3,300	3,700–5,500	4,600–7,250	5,500–9,000
1⅛–8		3,300–4,000	5,500–6,500	6,000–8,650	6,500–10,800
	1⅛–12	3,000–4,200	5,000–7,000	6,000–10,250	7,000–13,500
1¼–8		4,000–5,000	6,500–8,000	7,250–11,000	8,000–14,000
	1¼–12	5,400–6,600	9,000–11,000	10,000–16,750	11,000–22,500

Figure 6-13   (a) Standard torque table.

the holes. A thicker or extra washer might be required to correctly line up the holes. The torque loads specified may be used for all unlubricated cadmium-plated steel nuts of the fine or coarse thread series which have approximately equal number of threads

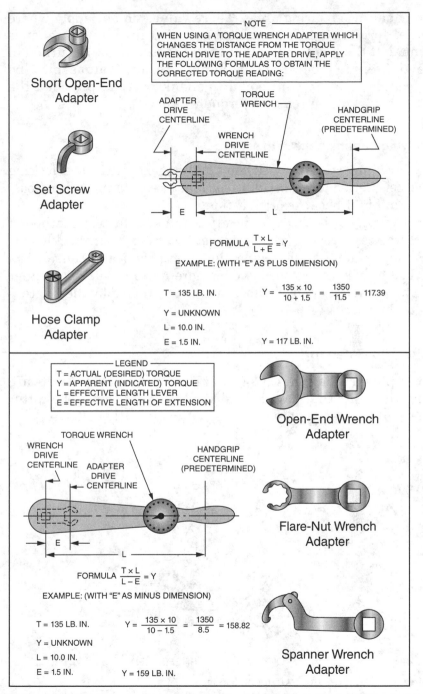

Short Open-End Adapter

Set Screw Adapter

Hose Clamp Adapter

NOTE

WHEN USING A TORQUE WRENCH ADAPTER WHICH CHANGES THE DISTANCE FROM THE TORQUE WRENCH DRIVE TO THE ADAPTER DRIVE, APPLY THE FOLLOWING FORMULAS TO OBTAIN THE CORRECTED TORQUE READING:

ADAPTER DRIVE CENTERLINE

TORQUE WRENCH

WRENCH DRIVE CENTERLINE

HANDGRIP CENTERLINE (PREDETERMINED)

E   L

FORMULA $\dfrac{T \times L}{L + E} = Y$

EXAMPLE: (WITH "E" AS PLUS DIMENSION)

T = 135 LB. IN.

$Y = \dfrac{135 \times 10}{10 + 1.5} = \dfrac{1350}{11.5} = 117.39$

Y = UNKNOWN
L = 10.0 IN.
E = 1.5 IN.

Y = 117 LB. IN.

LEGEND

T = ACTUAL (DESIRED) TORQUE
Y = APPARENT (INDICATED) TORQUE
L = EFFECTIVE LENGTH LEVER
E = EFFECTIVE LENGTH OF EXTENSION

TORQUE WRENCH

WRENCH DRIVE CENTERLINE   ADAPTER DRIVE CENTERLINE

HANDGRIP CENTERLINE (PREDETERMINED)

E   L

FORMULA $\dfrac{T \times L}{L - E} = Y$

EXAMPLE: (WITH "E" AS MINUS DIMENSION)

T = 135 LB. IN.

$Y = \dfrac{135 \times 10}{10 - 1.5} = \dfrac{1350}{8.5} = 158.82$

Y = UNKNOWN
L = 10.0 IN.
E = 1.5 IN.

Y = 159 LB. IN.

Open-End Wrench Adapter

Flare-Nut Wrench Adapter

Spanner Wrench Adapter

Figure 6-13 (b) Adjusting toque value if extensions are used with torque wrench.

and equal face-bearing areas. These values do not apply where special torque requirements are specified in the maintenance manual. If the head end, rather than the nut, must be turned in the tightening operation, maximum torque values may be increased by an amount equal to shank friction, provided the latter is first measured by a torque wrench.

Safetying of nuts, bolts, and screws

It is very important that all bolts or nuts, except the self-locking type, be safetied after installation. This prevents them from loosening in flight because of vibration.

Safety wiring is the most positive and satisfactory method of safetying capscrews, studs, nuts, boltheads, and turnbuckle barrels that cannot be safetied by any other practical means. It is a method of wiring together two or more units in such a manner that any tendency of one to loosen is counteracted by the tightening of the wire (Fig. 6-14).

Cotter Pin Safetying

Cotter pin installation is shown in Fig. 6-14. Castellated nuts are used with bolts that have been drilled for cotter pins. The cotter pin should fit neatly into the hole, with very little sideplay.

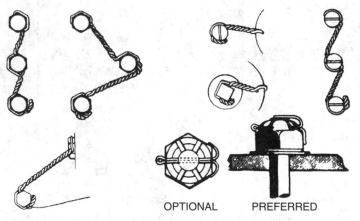

OPTIONAL   PREFERRED

Figure 6-14 Typical safety wiring methods.

Installation: Bolts, Washers, Nuts, and Cotter Pins

Use Fig. 6-15 as a guide to match all components of a bolted assembly.

Miscellaneous Threaded Fasteners

As stated earlier in this chapter, standard hex, slotted, and Phillips head-threaded fasteners are seldom used for structural applications on high-performance aircraft. For example, most threaded fasteners on the L-1011 jet transport aircraft are "Tri-Wing," developed by the Phillips Screw Company. Other types in general use are "Torq-Set" and "Hi-Torque®." All of these patented fasteners require special driving bits that fit into standard holders and screwdriver handles.

The Tri-Wing is shown in Fig. 6-16. Other fastener wrenching recesses are shown in Fig. 6-1. Various fasteners are illustrated in Chap. 13, Standard Parts.

BOLT		WASHER	NUT	COTTER PIN	
AN	DIAM.-THRD.	AN	AN	AN	DIAM.
−3	($^3/_{16}$) 10-32		310-3	380-2	$^1/_{16}$
−3A		960-10	365-1032	None	
4	$^1/_4$-28		310-4	380-2	$^1/_{16}$
4A		960-416	365-428	None	
5	$^5/_{16}$-24		310-5	380-2	$^1/_{16}$
5A		960-516	365-524	None	
6	$^3/_8$-24		310-6	380-3	$^3/_{32}$
6A		960-616	365-624	None	
7	$^7/_{16}$-20		310-7	380-3	$^3/_{32}$
7A		960-716	365-720	None	
8	$^1/_2$-20		310-8	380-3	$^3/_{32}$
8A		960-816	365-820	None	

Figure 6-15   Guide for installation of bolt, washer, nut, and cotter-pin assembly.

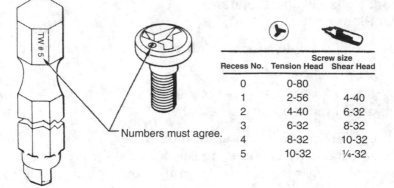

Recess No.	Screw size Tension Head	Shear Head
0	0-80	
1	2-56	4-40
2	4-40	6-32
3	6-32	8-32
4	8-32	10-32
5	10-32	¼-32

Numbers must agree.

Figure 6-16 Tri-wing heads are numbered for easy identification, and must be fitted with a similarly numbered bit for effective driving.

Screws

Screws are the most commonly used threaded fastening devices on aircraft. They differ from bolts in as much as they are generally made of lower strength materials. They can be installed with a loose-fitting thread, and the head shapes are made to engage a screwdriver or wrench. Some screws have a clearly defined grip or unthreaded portion while others are threaded along their entire length. Several types of structural screws differ from the standard structural bolts only in head style. The material in them is the same, and a definite grip length is provided. The AN525 washer head screw and the NAS220 through NAS227 series are such screws.

Commonly used screws are classified in three groups: (1) structural screws, which have the same strength as equal-size bolts; (2) machine screws, which include the majority of types used for general repair; and (3) self-tapping screws, which are used for attaching lighter parts, see Fig. 6-17.

Structural screws. Structural screws are made of alloy steel, are properly heat treated, and can be used as structural bolts. These screws are found in the NAS204 through NAS235 and AN509 and AN525 series. They have a definite grip and the same shear strength as a bolt of the same size. Shank tolerances are similar to AN hex-head bolts, and the threads are National Fine. Structural screws are available with round, brazier, or countersunk heads. The recessed head screws are driven by either a

Figure 6-17 Screws from left to right: structural screw, machine screw, and self-tapping screw.

Phillips or a Reed & Prince screwdriver. The AN509 (100-degree) flathead screw is used in countersunk holes where a flush surface is necessary. The AN525 washer head structural screw is used where raised heads are not objectionable. The washer head screw provides a large contact area.

Machine screws. Machine screws are usually of the flathead (countersunk), roundhead, or washer head types. These are general purpose screws and are available in low-carbon steel, brass, corrosion-resistant steel, and aluminum alloy. Roundhead screws, AN515 and AN520, have either slotted or recessed heads. The AN515 screw has coarse threads, and the AN520 has fine threads. Countersunk machine screws are listed as AN505 and AN510 for 82 degrees, and AN507 for 100 degrees. The AN505 and AN510 correspond to the AN515 and AN520 roundhead in material and usage. The fillister head screw, AN500 through AN503, is a general purpose screw and is used as a capscrew in light mechanisms. This could include attachments of cast aluminum parts such as gearbox cover plates. The AN500 and AN501 screws are available in low-carbon steel, corrosion-resistant steel, and brass. The AN500 has coarse threads, while the AN501 has fine threads. They have no clearly defined grip length. Screws larger than No. 6 have a hole drilled through the head for safetying purposes. The AN502 and AN503 fillister head screws are made of heat-treated alloy steel, have a small grip, and are available in fine and coarse threads. These screws are used as capscrews where great strength is required. The coarse-threaded screws are commonly used as capscrews in tapped aluminum alloy and magnesium castings because of the softness of the metal.

Self-tapping screws.   Self-tapping screws are listed as AN504 and AN506. The AN504 screw has a roundhead, and the AN506 is 82 degree countersunk. These screws are used for attaching removable parts, such as nameplates, to castings and parts in which the screw cuts its own threads.

AN530 and AN531 self-tapping sheet-metal screws, such as the Parker-Kalon Z-type sheet metal screw, are blunt on the end. They are used in the temporary attachment of metal for riveting, and in the permanent assembly of nonstructural assemblies. Self-tapping screws should not be used to replace standard screws, nuts, bolts, or rivets.

Dzus Fasteners

Although not a threaded fastener, the Dzus fastener is an example of a quick-disconnect fastener, such as used on a cowling or nacelle.

The Dzus turnlock fastener consists of a stud, grommet, and receptacle. Figure 6-18 illustrates an installed Dzus fastener and the various parts.

The grommet is made of aluminum or aluminum alloy material. It acts as a holding device for the stud. Grommets can be fabricated from 1100 aluminum tubing, if none are available from normal sources.

The spring is made of steel, cadmium-plated to prevent corrosion. The spring supplies the force that locks or secures the stud in place when two assembles are joined.

The studs are fabricated from steel and are cadmium-plated. They are available in three head styles: wing, flush, and oval.

A quarter of a turn of the stud (clockwise) locks the fastener. The fastener can be unlocked only by turning the stud counterclockwise. A Dzus key (or a specially ground screwdriver) locks

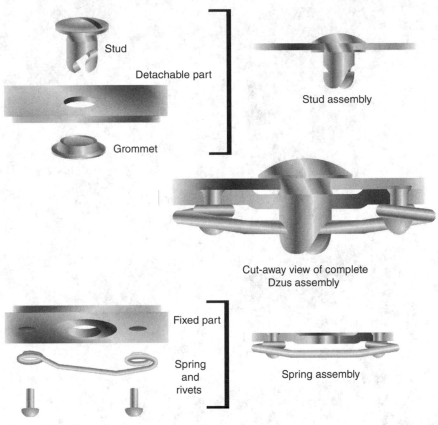

Figure 6-18 Dzus fastener.

or unlocks the fastener. Special installation tools and instructions are available from the manufacturers.

Camloc Fasteners

Camloc fasteners are made in a variety of styles and designs. Included among the most commonly used are the 2600, 2700, 40S51, and 4002 series in the regular line, and the stressed panel fastener in the heavy duty line. The latter is used in stressed panels which carry structural loads. The Camloc fastener is used to secure aircraft cowlings and fairings. It consists of three parts: a stud assembly, a grommet, and a receptacle. Two types of receptacles are available: rigid and floating as shown in Fig. 6-19.

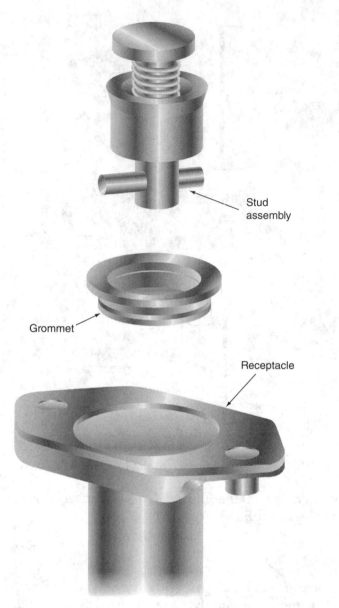

Figure 6-19 Camloc fastener.

The stud and grommet are installed in the removable portion; the receptacle is riveted to the structure of the aircraft. The stud and grommet are installed in either a plain, dimpled, countersunk, or counterbored hole, depending upon the location and thickness of the material involved. A quarter turn (clockwise) of the stud locks the fastener. The fastener can be unlocked only by turning the stud counterclockwise.

Hi-Lok®, Hi-Tigue®, and Hi-Lite® Fasteners

Hi-Lok® fastening system

The threaded end of the Hi-Lok® two-piece fastener contains a hexagonal-shaped recess as shown in Fig. 6-20. The hex tip of an Allen wrench engages the recess to prevent rotation of the pin while the collar is being installed (Fig. 6-21). The pin is designed in two basic head styles. For shear applications, the pin is made in countersunk style and in a compact protruding head style. For tension applications, the MS24694 countersunk and regular protruding head styles are available. The self-locking, threaded Hi-Lok® collar has an internal counterbore at the base to accommodate variations in material thickness. At the opposite end of the collar is a wrenching device that is torqued by the driving tool until it shears off during installation, leaving the lower portion of the collar seated with the proper torque without additional torque inspection. This shear-off point occurs when a predetermined preload or clamp-up is attained in the fastener during installation.

The advantages of Hi-Lok® two-piece fastener include its light weight, high fatigue resistance, high strength, and its inability to be overtorqued. The pins, made from alloy steel, corrosion-resistant steel, or titanium alloy, come in many standard and oversized shank diameters. The collars are made of aluminum alloy, corrosion-resistant steel, or alloy steel. The collars have

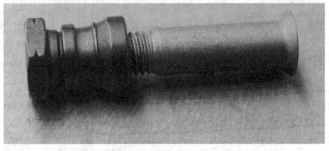

Figure 6-20 Hi-Lok® fastener and collar.

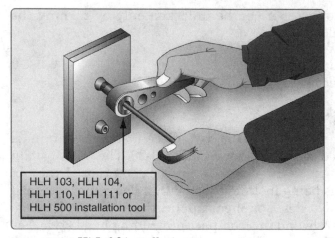

Figure 6-21   Hi-Lok® installation.

wrenching flats, fracture point, threads, and a recess. The wrenching flats are used to install the collar. The fracture point has been designed to allow the wrenching flats to shear when the proper torque has been reached. The threads match the threads of the pins and have been formed into an ellipse that is distorted to provide the locking action. The recess serves as a built-in washer. This area contains a portion of the shank and the transition area of the fastener. The hole shall be prepared so that the maximum interference fit does not exceed 0.002 inch. This avoids build up of excessive internal stresses in the work adjacent to the hole.

The Hi-Lok® pin has a slight radius under its head to increase fatigue life. After drilling, deburr the edge of the hole to allow the head to seat fully in the hole. The Hi-Lok® is installed in interference fit holes for aluminum structure and a clearance fit for steel, titanium, and composite materials. Figure 6-21 shows an example of a Hi-Lok® basic partnumber.

Hi-Tigue® fastening system

The Hi-Tigue® fastener offers all of the benefits of the Hi-Lok® fastening system along with a unique bead design that enhances the fatigue performance of the structure making it ideal for situations that require a controlled interference fit. The Hi-Tigue® fastener assembly consists of a pin and collar. These pin rivets have a radius at the transition area. During installation in an interference fit hole, the radius area will "cold work" the hole. These fastening systems can be easily confused, and visual reference should not be used for identification. Use part numbers to identify these fasteners. Figure 6-22 shows the differences between a Hi-Lok® and a Hi-Tigue® pin.

Hi-Lite® fastening system

The Hi-Lite® fastener is similar in design and principle to the Hi-Lok® fastener, but the Hi-Lite® fastener has a shorter transition area between the shank and the first load-bearing thread. Hi-Lite® has approximately one less thread. All Hi-Lite® fasteners are made of titanium. These differences reduce the weight of the Hi-Lite® fastener without lessening the shear strength, but the Hi-Lite® clamping forces are less than that of a Hi-Lok® fastener. The Hi-Lite® collars are also different and thus are not interchangeable with Hi-Lok® collars. Hi-Lite® fasteners can be replaced with Hi-Lok® fasteners for most applications, but Hi-Loks® cannot be replaced with Hi-Lites®.

Installation of Hi-Lok®, Hi-Tigue®, and Hi-Lite® Fasteners

Hole preparation

Hi-Lok® pins require reamed and chamfered holes, and, in some cases, an interference fit. For standard Hi-Lok® pins, it is generally recommended that the maximum interference fit shall not exceed 0.002 inch. The Hi-Tigue®-type Hi-Lok® pin is normally installed in a hole with a 0.002- to 0.004-inch diametral interference.

The Hi-Lok® pin has a slight radius under its head (Fig. 6-22). After drilling, deburr the edge of the hole. This permits the head

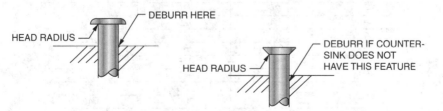
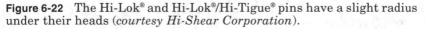

Figure 6-22 The Hi-Lok® and Hi-Lok®/Hi-Tigue® pins have a slight radius under their heads (*courtesy Hi-Shear Corporation*).

to fully seat in the hole. See the appropriate Hi-Lok® standards for head radius dimensions. For example, the ³⁄₁₆ protruding head has a 0.015/0.025 radius, and the ³⁄₁₆ flush head has a 0.025/0.030 radius.

Pin grip length

Standard pin lengths are graduated in ¹⁄₁₆ inch increments. The material thickness can vary ¹⁄₁₆ inch without changing pin lengths. Adjustment for variations in material thickness in between the pin ¹⁄₁₆ inch graduations is automatically made by the counterbore in the collar (Fig. 6-23). The grip length is determined, as shown in Fig. 6-24.

Installation tools

Hi-Lok® fasteners are rapidly installed by one person working from one side of the work using standard power or hand tools and Hi-Lok® adaptor tools.

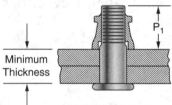
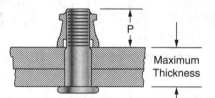

MINIMUM GRIP
(Maximum Protrusion)

MAXIMUM GRIP
(Minimum Protrusion)

Standard Hi-Lok Pin		Minimum Protrusion P	Maximum Protrusion P1
First Dash Number	Nominal Diameter		
–5	5/32	.302	.384
–6	3/16	.315	.397
–8	1/4	.385	.467
–10	5/16	.490	.572
–12	3/8	.535	.617
–14	7/16	.625	.707
–16	1/2	.675	.757
–18	9/16	.760	.842
–20	5/8	.815	.897
–24	3/4	1.040	1.122
–28	7/8	1.200	1.282
–32	1	1.380	1.462

Figure 6-23 Table showing installed Hi-Lok® pin protrusion limits (*courtesy Hi-Shear Corporation*).

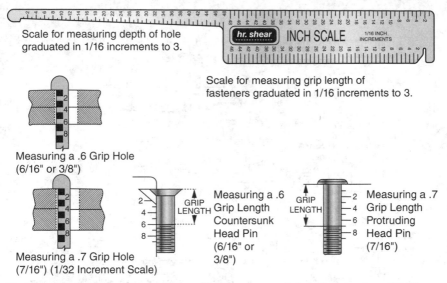

Scale for measuring depth of hole graduated in 1/16 increments to 3.

hr. shear INCH SCALE 1/16 INCH INCREMENTS

Scale for measuring grip length of fasteners graduated in 1/16 increments to 3.

Measuring a .6 Grip Hole (6/16" or 3/8")

Measuring a .7 Grip Hole (7/16") (1/32 Increment Scale)

GRIP LENGTH Measuring a .6 Grip Length Countersunk Head Pin (6/16" or 3/8")

GRIP LENGTH Measuring a .7 Grip Length Protruding Head Pin (7/16")

Figure 6-24 Determining grip length using a special scale (*courtesy Hi-Shear Corporation*).

Hi-Lok® adaptor tools are fitted to high-speed pistol grip and ratchet wrench drives in straight, 90 degree, offset extension, and automatic collar-feed configurations. Figure 6-25 shows a few of the hand and power tools available for installing Hi-Lok® and Hi-Lok®/Hi-Tigue® fasteners.

The basic consideration in determining the correct hand tool is to match the socket-hex tip dimensions of the tool with the Hi-Lok®/Hi-Tigue® pin hex recess and collar-driving hex of the particular pin-collar combination to be installed. Figure 6-26 indicates the hex dimensions that must match.

Installation steps for an interference-fit hole

Figure 6-27 shows the installation steps in a noninterference-bit hole. When Hi-Lok®/Hi-Tigues® are installed in an interference-fit, the pins should be driven in using a standard rivet gun and Hi-Tigue® pin driver, as shown in Fig. 6-28. The structure must be supported with a draw bar, as shown.

When Hi-Lok®/Hi-Tigue® pins are pressed or tapped into holes, the fit is sufficiently tight to grip the pin to prevent it from rotating.

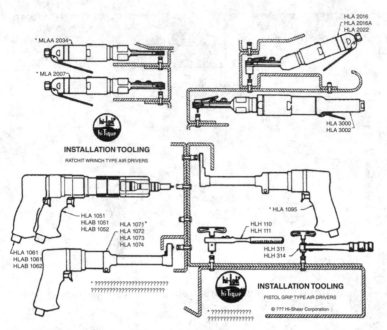

Figure 6-25 A few of the hand and power tools available for installing Hi-Lok® and Hi-Lok®/Hi-Tigue® fasteners (*courtesy Hi-Shear Corporation*).

Hi-Lok® driver tools are available that use a finder pin, instead of the hex wrench tip to locate the tool on the collar and pin (Fig. 6-29). Otherwise, installation steps for interference-fit holes are the same as for standard Hi-Lok® fasteners.

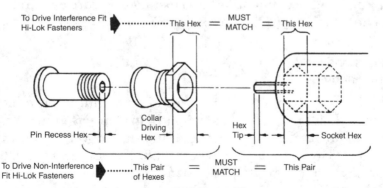

Figure 6-26 Determining the correct hand tool by matching hex dimensions (*courtesy Hi-Shear Corporation*).

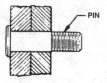

a. Insert the pin into the prepared non-interference fit hole.

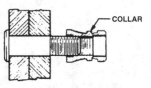

b. Manually thread the collar onto the pin.

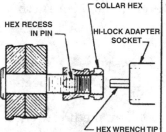

c. Insert the hex wrench tip of the power driver into the pin's hex recess, and the socket over the collar hex. This prevents rotation of the pin while the collar is being installed.

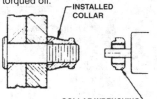

d. Firmly press the power driver against the collar, operate the power driver until the collar's wrenching device has been torqued off.

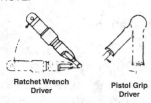

e. This completes the installation of the Hi-Lok Fastener Assembly.

NOTE:

Ratchet Wrench Driver

Pistol Grip Driver

To ease the removal of the driving tool's hex wrench tip from the hex recess of the pin after the collar's wrenching device has sheared off, simply rotate the entire driver tool in a slight clockwise motion.

Figure 6-27 Installation steps in noninterference fit hole (*courtesy Hi-Shear Corporation*).

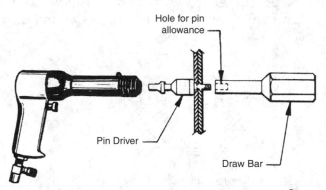

Figure 6-28 Installing an interference fit Hi-Tigue® pin using a rivet gun (*courtesy Hi-Shear Corporation*).

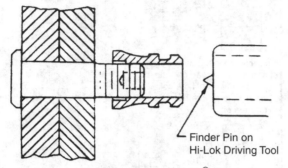

Figure 6-29 Finder pin on Hi-Lok® driving tool (*courtesy Hi-Shear Corporation*).

For field service, all sizes of Hi-Lok® fasteners can be installed with hand tools (standard Allen hex keys and open-end or ratchet-type wrenches).

Inspection after installation

Hi-Lok® and Hi-Lok®/Hi-Tigue® fasteners are visually inspected. No torque wrenches are required.

The Hi-Lok® protrusion gauges offer a convenient method to check Hi-Lok® pin-protrusion limits after the Hi-Lok® pin has been inserted in the hole and before or after collar installation (Fig. 6-30). Individual gauges accommodate Hi-Lok® pin diameter sizes of $\frac{5}{32}$ inch, $\frac{3}{16}$ inch, $\frac{1}{4}$ inch, $\frac{5}{16}$ inch, and $\frac{3}{8}$ inch. Gauges are made of 0.012 inch stainless steel and are assembled as a set on a key chain.

Removal of the installed fastener

Removal of fasteners is accomplished with standard hand tools in a manner similar to removing a nut from a bolt. By holding the

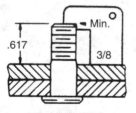

MINIMUM GRIP
(Maximum Protrusion)

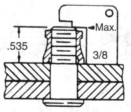

MAXIMUM GRIP
(Minimum Protrusion)

Figure 6-30 Protrusion limits for standard Hi-Lok® pins; $\frac{3}{8}$ gauge is shown as an example (*courtesy Hi-Shear Corporation*).

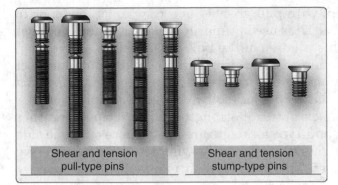

Figure 6-31   Shear and tension lockbolts.

pin with a standard Allen wrench, the collar can be removed with pliers. Hollow mill-type cutters attached to power tools can also remove the collars without damage to the pin, and the pins can be reused if they are undamaged. Special hand and power removal tools are also available.

Lockbolt Fastening Systems

The lockbolt is a two-piece fastener that combines the features of a high-strength bolt and a rivet with advantages over each (Fig. 6-31). In general, a lockbolt is a nonexpanding fastener that has either a collar swaged into annular locking groves on the pin shank or a type of threaded collar to lock it in place. Available with either countersunk or protruding heads, lockbolts are permanent type fasteners assemblies and consist of a pin and a collar.

Often called huckbolts, lockbolts are manufactured by companies such as Cherry® Aerospace (Cherry® Lockbolt), Alcoa Fastening Systems (Hucktite® Lockbolt System), and SPS Technologies. Used primarily for heavily stressed structures that require higher shear and clamp-up values than can be obtained with rivets. Three types of lockbolts are commonly used: pull-type, stump-type, and blind-type.

The pull-type lockbolt is mainly used in aircraft primary and secondary structure. It is installed very rapidly and has approximately one-half the weight of equivalent AN steel bolts and nuts. A special pneumatic pull gun is required for installation of this type lockbolt, which can be performed by one operator since buckling is not required. The stump-type lockbolt, although not having the extended stem with pull grooves, is a companion fastener to the pulltype lockbolt. It is used primarily where clearance does not permit effective installation of the pull-type lockbolt. It is driven with a standard pneumatic riveting hammer, with a hammer set attached for swaging the collar into the pin locking grooves, and a bucking bar. The blind-type lockbolt comes as a complete unit or assembly and has exceptional strength and sheet pull-together characteristics. Blind-type lockbolts are used where only one side of the work is accessible and generally where it is difficult to drive a conventional rivet. This type lockbolt is installed in a manner similar to the pull-type lockbolt.

Installation procedure

Installation of lockbolts involves proper drilling. The hole preparation for a lockbolt is similar to hole preparation for a Hi-Lok®. An interference fit is typically used for aluminum and a clearance fit is used for steel, titanium, and composite materials. A pneumatic pull gun is used as shown in Fig. 6-32.

Lockbolt inspection

After installation, a lockbolt needs to be inspected to determine if installation is satisfactory as shown in Fig. 6-33.

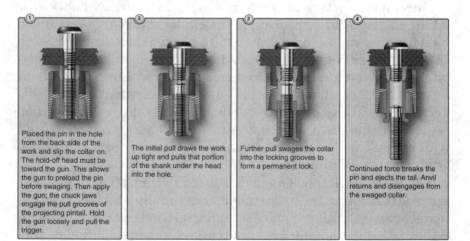

① Placed the pin in the hole from the back side of the work and slip the collar on. The hold-off head must be toward the gun. This allows the gun to preload the pin before swaging. Then apply the gun; the chuck jaws engage the pull grooves of the projecting pintail. Hold the gun loosely and pull the trigger.

② The initial pull draws the work up tight and pulls that portion of the shank under the head into the hole.

③ Further pull swages the collar into the locking grooves to form a permanent lock.

④ Continued force breaks the pin and ejects the tail. Anvil returns and disengages from the swaged collar.

Figure 6-32 Installation of pull-type lockbolt.

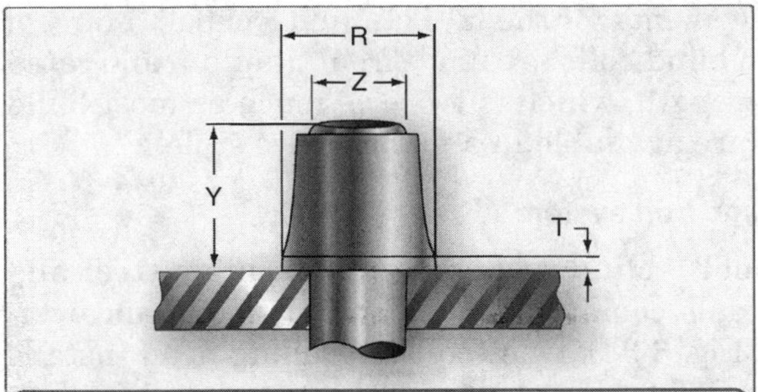

Lockbolt/Collar Acceptance Criteria				
Nominal Fastener Diameter	Y	Z (Ref.)	R Max.	T Min.
5/32	.324/.161	.136	.253	.037
3/16	.280/.208	.164	.303	.039
1/4	.374/.295	.224	.400	.037
5/16	.492/.404	.268	.473	.110
3/8	.604/.507	.039	.576	.120

Figure 6-33 Lockbolt inspection.

Inspect the lockbolt as follows:

1. The head must be firmly seated.
2. The collar must be tight against the material and have the proper shape and size.
3. Pin protrusion must be within limits.

Lockbolt removal

The best way to remove a lockbolt is to remove the collar and drive out the pin. The collar can be removed with a special collar cutter attached to a drill motor that mills off the collar without damaging the skin. If this is not possible, a collar splitter or small chisel can be used. Use a backup block on the opposite side to prevent elongation of the hole.

Blind Bolts

Blind bolts have a higher strength than blind rivets and are used for joints that require high strength. Sometimes, these bolts can be direct replacements for the Hi-Lok® and lockbolt. Many of the new generation blind bolts are made from titanium and rated at 90 KSI shear strength, which is twice as much as most blind rivets. Blind bolts are available in a pull or drive style.

Cherry Maxibolt® blind bolt system

The Cherry Maxibolt® blind bolt, available in alloy steel and A-286 CRES materials, comes in four different nominal and over-sized head styles (Fig. 6-34). One tool and pulling head installs all three diameters. The blind bolts create a larger blind side footprint and they provide excellent performance in thin sheet and nonmetallic applications. The flush breaking stem eliminates shaving while the extended grip range accommodates different application thicknesses.

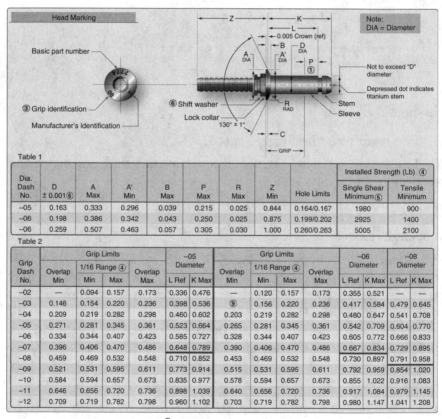

Table 1

Dia. Dash No.	D ± 0.001 ⑥	A Max	A' Min	B Max	P Max	R Max	Z Min	Hole Limits	Installed Strength (Lb) ④ Single Shear Minimum ⑤	Tensile Minimum
−05	0.163	0.333	0.296	0.039	0.215	0.025	0.844	0.164/0.167	1980	900
−06	0.198	0.386	0.342	0.043	0.250	0.025	0.875	0.199/0.202	2925	1400
−06	0.259	0.507	0.463	0.057	0.305	0.030	1.000	0.260/0.263	5005	2100

Table 2

Grip Dash No.	Grip Limits Overlap Min	1/16 Range ④ Min	1/16 Range ④ Max	Overlap Max	−05 Diameter L Ref	−05 Diameter K Max	Grip Limits Overlap Min	1/16 Range ④ Min	1/16 Range ④ Max	Overlap Max	−06 Diameter L Ref	−06 Diameter K Max	−08 Diameter L Ref	−08 Diameter K Max
−02	—	0.094	0.157	0.173	0.336	0.476	—	0.120	0.157	0.173	0.355	0.521	—	—
−03	0.146	0.154	0.220	0.236	0.398	0.536	⑨	0.156	0.220	0.236	0.417	0.584	0.479	0.645
−04	0.209	0.219	0.282	0.298	0.460	0.602	0.203	0.219	0.282	0.298	0.480	0.647	0.541	0.708
−05	0.271	0.281	0.345	0.361	0.523	0.664	0.265	0.281	0.345	0.361	0.542	0.709	0.604	0.770
−06	0.334	0.344	0.407	0.423	0.585	0.727	0.328	0.344	0.407	0.423	0.605	0.772	0.666	0.833
−07	0.396	0.406	0.470	0.486	0.648	0.789	0.390	0.406	0.470	0.486	0.667	0.834	0.729	0.895
−08	0.459	0.469	0.532	0.548	0.710	0.852	0.453	0.469	0.532	0.548	0.730	0.897	0.791	0.958
−09	0.521	0.531	0.595	0.611	0.773	0.914	0.515	0.531	0.595	0.611	0.792	0.959	0.854	1.020
−10	0.584	0.594	0.657	0.673	0.835	0.977	0.578	0.594	0.657	0.673	0.855	1.022	0.916	1.083
−11	0.646	0.656	0.720	0.736	0.898	1.039	0.640	0.656	0.720	0.736	0.917	1.084	0.979	1.145
−12	0.709	0.719	0.782	0.798	0.960	1.102	0.703	0.719	0.782	0.798	0.980	1.147	1.041	1.208

Figure 6-34 Cherry Maxibolt® installation information.

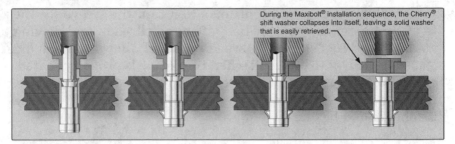

During the Maxibolt® installation sequence, the Cherry® shift washer collapses into itself, leaving a solid washer that is easily retrieved.

Figure 6-35 Cherry Maxibolt® installation procedure.

Cherry Maxibolts® are primarily used in structures where higher loads are required. The steel version is 112 KSI shear. The A286 version is 95 KSI shear. The Cherry® G83, G84, or G704 installation tools are required for installation. Figure 6-35 shows the installation procedure.

Drive-nut-type blind bolt

Jo-bolts, Visu-Lok®, Composi-Lok®, OSI Bolt®, and Radial-Lok® fasteners use the drive nut concept and are composed of a nut, sleeve, and a draw bolt as shown in Fig. 6-36. These types of blind bolts are used for high-strength applications in metals and composites when there is no access to the blind side. Available in steel and titanium alloys, they are installed with special tooling. Both powered and hand tooling is available. During installation, the nut is held stationary while the core bolt is rotated by the installation tooling. The rotation of the core bolt draws the sleeve into

Figure 6-36 Drive-nut-type blind bolt.

the installed position and continues to retain the sleeve for the life of the fastener. The bolt has left-hand threads and driving flats on the threaded end. A break-off relief allows the driving portion of the bolt to break off when the sleeve is properly seated. These types of bolts are available in many different head styles, including protruding head, 100-degree flush head, 130-degree flush head, and hex head.

7

Aircraft Plumbing

Fluid Lines

Aircraft plumbing lines usually are made of metal tubing and fittings or of flexible hose. Metal tubing is widely used in aircraft for fuel, oil, coolant, oxygen, instrument, and hydraulic lines. Flexible hose is generally used with moving parts or where the hose is subject to considerable vibration.

In modern aircraft, aluminum alloy, corrosion-resistant steel or titanium tubing have generally replaced copper tubing.

The workability, resistance to corrosion, and light weight of aluminum alloy are major factors in its adoption for aircraft plumbing.

Aluminum alloy tubing

Tubing made from 1100 H14 (½-hard) or 3003 H14 (½-hard) is used for general purpose lines of low or negligible fluid pressures, such as instrument lines and ventilating conduits. Tubing made from 2024-T3, 5052-O, and 6061-T6 aluminum alloy materials is used in general purpose systems of low and medium pressures, such as hydraulic and pneumatic 1000 to 1500 psi systems, and fuel and oil lines.

Steel

Corrosion-resistant steel tubing, either annealed CRES 304, CRES 321, or CRES 304-1/8-hard, is used extensively in high-pressure hydraulic systems (3000 psi or more) for the operation of landing gear, flaps, brakes, and in fire zones. Its higher tensile strength permits the use of tubing with thinner walls; consequently, the final installation weight is not much greater than that of the thicker wall aluminum alloy tubing. Steel lines are used where there is a risk of foreign object damage (FOD); that is the landing gear and wheel well areas. Although identification markings for steel tubing differ, each usually includes the manufacturer's name or trademark, the *Society of Automotive Engineers* (SAE) number, and the physical condition of the metal.

Titanium 3AL-2.5V

This type of tubing and fitting is used extensively in transport category and high-performance aircraft hydraulic systems for pressures above 1500 psi. Titanium is 30 percent stronger than steel and 50 percent lighter than steel. Cryofit fittings or swaged fittings are used with titanium tubing. Do not use titanium tubing and fittings in any oxygen system assembly. Titanium and titanium alloys are oxygen reactive. If a freshly formed titanium surface is exposed in gaseous oxygen, spontaneous combustion could occur at low pressures.

Tubing identification

Aluminum alloy, steel, or titanium tubing can be identified readily by sight where it is used as the basic tubing material. However, it is difficult to determine whether a material is carbon steel or stainless steel, or whether it is 1100, 3003, 5052-O, 6061-T6, or 2024-T3 aluminum alloy. To positively identify the material used in the original installation, compare code markings of the replacement tubing with the original markings on the tubing being replaced. On large aluminum alloy tubing, the alloy designation is stamped on the surface. On small aluminum tubing, the designation may be stamped on the surface; but more often it is shown by a color code, not more than 4 inch in width, painted at the two ends and approximately midway between the ends of some tubing. When the band consists of two colors, one-half the width is used for each color. Figure 7-1 shows the color coding for aluminum tubing.

Aluminum Alloy Number	Color of Band
1100	White
3003	Green
2014	Gray
2024	Red
5052	Purple
6053	Black
6061	Blue and Yellow
7075	Brown and Yellow

Figure 7-1 Aluminum tubing identification codes.

Sizes

Metal tubing is sized by outside diameter (o.d.), which is measured fractionally in sixteenths of an inch. Thus, number 6 tubing is $\frac{6}{16}$ inch (or $\frac{3}{8}$ inch) and number 8 tubing is $\frac{8}{16}$ inch (or $\frac{1}{2}$ inch), and so forth. The tube diameter is typically printed on all rigid tubing. In addition to other classifications or means of identification, tubing is manufactured in various wall thicknesses. Thus, it is important when installing tubing to know not only the material and outside diameter, but also the thickness of the wall. The wall thickness is typically printed on the tubing in thousands of an inch. To determine the inside diameter (i.d.) of the tube, subtract twice the wall thickness from the outside diameter. For example, a number 10 piece of tubing with a wall thickness of 0.063 inch has an inside diameter of 0.625 inch − 2(0.063 inch) = 0.499 inch.

Flexible Hose

Flexible hose is used in aircraft plumbing to connect moving parts with stationary parts in locations subject to vibration or where a great amount of flexibility is needed. It can also sense a connector in metal tubing systems.

Synthetics

Synthetic materials most commonly used in the manufacture of flexible hose are Buna-N, Neoprene, Butyl, and Teflon. Buna-N is a synthetic rubber compound that has excellent resistance to petroleum products. Do not confuse with Buna-S. Do not use for phosphate ester-based hydraulic fluid (Skydrol). Neoprene is a synthetic rubber compound that has an acetylene base. Its resistance to petroleum products is not as good as Buna-N, but it has better abrasive resistance. Do not use for phosphate ester-based hydraulic fluid (Skydrol). Butyl is a synthetic rubber compound made from petroleum raw materials. It is an excellent material to use with phosphate ester-based hydraulic fluid (Skydrol). Do not use it with petroleum products. *Teflon* is the DuPont trade name for tetra-fluorethylene resin. It has a broad operating temperature range (–65°F to 450°F). It is compatible with nearly every substance or agent used. It offers little resistance to flow; sticky viscous materials will not adhere to it. It has less volumetric expansion than rubber and the shelf and service life is practically limitless.

Rubber hose

Flexible rubber hose consists of a seamless synthetic rubber inner tube covered with layers of cotton braid and wire braid, and an outer layer of rubber-impregnated cotton braid. This type of hose is suitable for use in fuel, oil, coolant, and hydraulic systems. The types of hose are normally classified by the amount of pressure they are designed to withstand under normal operating conditions:

- Low pressure; any pressure below 250 psi, and fabric braid reinforcement.
- Medium pressure; pressures up to 3000 psi, and one wire braid reinforcement. Smaller sizes carry pressure up to 3000 psi; larger sizes carry pressure up to 1000 psi.
- High pressure; all sizes up to 3000 psi operating pressures.

Teflon hose

Teflon hose is a flexible hose designed to meet the requirements of higher operating temperatures and pressures in present aircraft systems. It can generally be used in the same manner as rubber hose. Teflon hose is processed and extruded into tube shapes of a desired size. It is covered with stainless steel wire, which is braided over the tube for strength and protection.

Teflon hose is unaffected by any known fuel, petroleum, or synthetic-based oils, alcohol, coolants, or solvents commonly used in aircraft. Although it is highly resistant to vibration and fatigue, the principle advantage of this hose is its operating strength.

Identification of hose

Identification markings of lines, letters, and numbers are printed on the hose (Fig. 7-2). These code markings show such information as hose size, manufacturer, date of manufacture, and pressure and temperature limits. Code markings assist in replacing a hose with one of the same specification or a recommended substitute. A hose

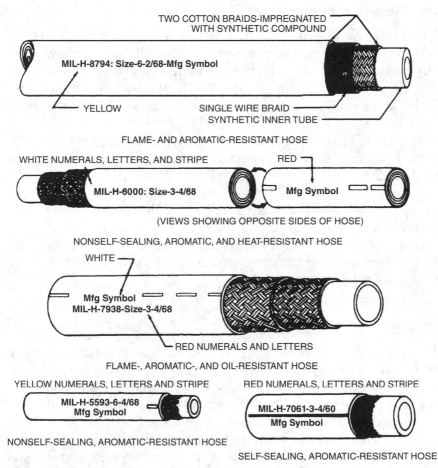

Figure 7-2 Hose-identification markings.

suitable for use with phosphate ester-based hydraulic fluid is marked "Skydrol use." In some instances, several types of hose might be suitable for the same use. Therefore, to make the correct hose selection, always refer to the maintenance or parts manual for the particular aircraft.

Size designation

The size of flexible hose is determined by its inside diameter. Sizes are in $\frac{1}{16}$ inch increments and are identical to corresponding sizes of rigid tubing, with which it can be used.

Identification of fluid lines

Fluid lines in aircraft are often identified by markers consisting of color codes, words, and geometric symbols. These markers identify each line's function, content, and primary hazard, as well as the direction of fluid flow. Figure 7-3 illustrates the various color codes and symbols used to designate the type of system and its contents.

In addition to the previously mentioned markings, certain lines can be further identified regarding specific function within a system: DRAIN, VENT, PRESSURE, or RETURN.

Generally, tapes and decals are placed on both ends of a line and at least once in each compartment through which the line runs. In addition, identification markers are placed immediately adjacent to each valve, regulator, filter, or other accessory within a line. Where paint or tags are used, location requirements are the same as for tapes and decals.

Plumbing Connections

Plumbing connectors, or fittings, attach one piece of tubing to another or to system units. The four types are: flared, flareless, bead and clamp, and swaged and welded. The beaded joint, which requires a bead and a section of hose and hose clamps, is used only in low- or medium-pressure systems, such as vacuum and coolant systems. The flared, flareless, and swaged types can be used as connectors in all systems, regardless of the pressure.

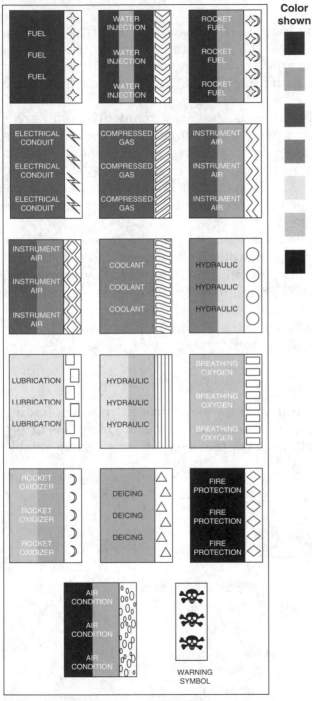

Figure 7-3 Identification of fluid lines.

Flared-tube fittings

A flared-tube fitting consists of a sleeve and a nut, as shown in Fig. 7-4. The nut fits over the sleeve and, when tightened, draws the sleeve and tubing flare tightly against a male fitting to form a seal. Tubing used with this type of fitting must be flared before installation.

The AN standard fitting is the most commonly used flared-tubing assembly for attaching the tubing to the various fittings required in aircraft plumbing systems. The AN standard fittings include the AN818 nut and AN819 sleeve. The AN819 sleeve is used with the AN818 coupling nut. All of these fittings have straight threads, but they have different pitch for the various types.

Flared-tube fittings are made of aluminum alloy, steel, or copper-based alloys. For identification purposes, all AN steel fittings are colored black and all AN aluminum alloy fittings are colored blue. The AN819 aluminum bronze sleeves are cadmium plated and are not colored. The size of these fittings is given in dash numbers, which equal the nominal tube outside diameter (O.D.) in sixteenths of an inch.

Flareless-tube fittings

The MS (military standard) flareless-tube fittings are finding wide application in aircraft plumbing systems. Using this fitting eliminates all tube flaring, yet provides safe, strong, dependable tube connections (Fig. 7-5).

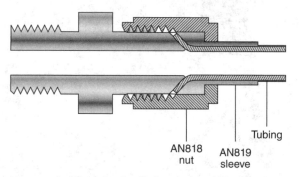

Figure 7-4   Flared tube fitting using AN parts.

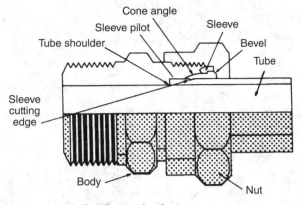

Figure 7-5 A flareless tube fitting.

Swaged fittings

A popular repair system for connecting and repairing hydraulic lines on transport category aircraft is the use of Permaswage™ fittings. Swaged fittings create a permanent connection that is virtually maintenance free. Swaged fittings are used to join hydraulic lines in areas where routine disconnections are not required and are often used with titanium and corrosion-resistant steel tubing. The fittings are installed with portable hydraulically powered tooling, which is compact enough to be used in tight spaces as shown in Fig. 7-6. If the fittings need to be disconnected, cut the tubing with a tube cutter. Special installation tooling is available in portable kits. Always use the manufacturer's instructions to install swaged fittings. One of the latest developments is the Permalite™ fitting. Permalite™ is a tube fitting that is mechanically attached to the tube by axial swaging. The movement of the ring along the fitting body results in deformation of the tube with a leak-tight joint.

Cryofit fittings

Many transport category aircraft use Cryofit fittings to join hydraulic lines in areas where routine disconnections are not required. Cryofit fittings are standard fittings with a cryogenic sleeve. The sleeve is made of a shape memory alloy, Tinel™. The sleeve is manufactured three percent smaller, frozen in liquid nitrogen, and expanded to five percent larger than the line.

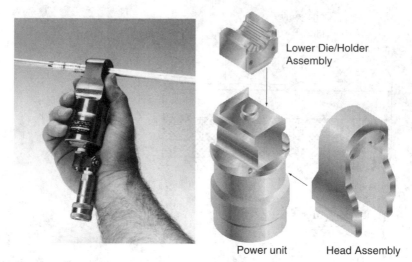

Lower Die/Holder Assembly

Power unit   Head Assembly

Figure 7-6 Installation of Permaswage™ fittings.

During installation, the fitting is removed from the liquid nitrogen and inserted onto the tube. During a 10- to 15-second warming up period, the fitting contracts to its original size (three percent smaller), biting down on the tube, forming a permanent seal. Cryofit fittings can only be removed by cutting the tube at the sleeve, though this leaves enough room to replace it with a swaged fitting without replacing the hydraulic line. It is frequently used with titanium tubing. The shape memory technology is also used for end fittings, flared fittings, and flareless fittings.

Tube cutting

When cutting tubing, it is important to produce a square end, free of burrs. Tubing can be cut with a tube cutter (Fig. 7-7) or a hacksaw. The cutter can be used with any soft metal tubing, such as copper, aluminum, or aluminum alloy.

Special chipless cutters are available for cutting aluminum 6061-T6, corrosion-resistant steel, and titanium tubing.

If a tube cutter is not available, or if hard material tubing is to be cut, use a fine-tooth hacksaw, preferably one having 32 teeth per inch. After sawing, file the end of the tube square and smooth, removing all burrs.

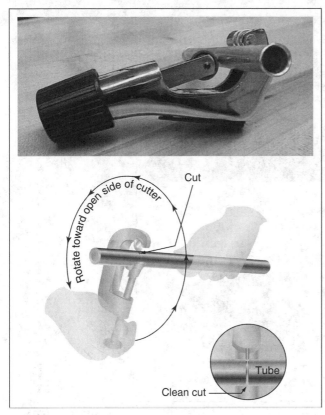

Figure 7-7   A hand-operated tube cutter.

Deburring

After cutting the tubing, carefully remove any burrs from inside and outside the tube. Use a knife or the burring edge attached to the tube cutter. The deburring operation can be accomplished by the use of a deburring tool as shown in Fig. 7-8. This tool is capable of removing both the inside and outside burrs by just turning the tool end for end. When performing the deburring operation, use extreme care that the wall thickness of the end of the tubing is not reduced or fractured. Very slight damage of this type can lead to fractured flares or defective flares which will not seal properly. Use a fine-tooth file to file the end square and smooth.

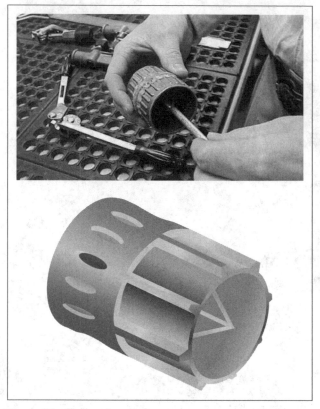

Figure 7-8   Deburring tool.

Tube bending

The objective in tube bending is to obtain a smooth bend without flattening the tube. Tubing less than ¾ inch in diameter usually can be bent with a hand bending tool (Fig. 7-9). For larger sizes, a factory tube-bending machine is usually used.

Tube-bending machines for all types of tubing are generally used in repair stations and large maintenance shops. With such equipment, proper bends can be made on large-diameter tubing and on tubing made from hard material. The production tube bender is one example.

Bend the tubing carefully to avoid excessive flattening, kinking, or wrinkling. A small amount of flattening in bends is acceptable, but the small diameter of the flattened portion must not be less than 75 percent of the original outside diameter. Tubing with

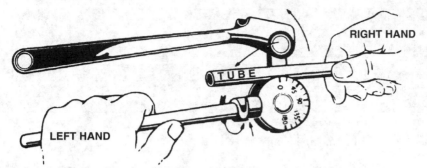

Figure 7-9   A hand tube bender.

flattened, wrinkled, or irregular bends should not be installed. Wrinkled bends usually result from trying to bend thin-wall tubing without using a tube bender. Examples of correct and incorrect tubing bends are shown in Fig. 7-10.

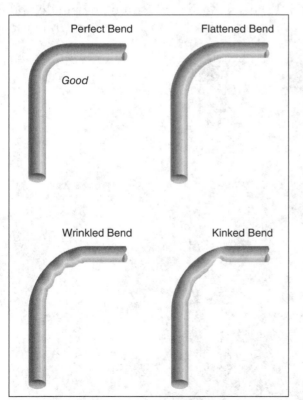

Figure 7-10   Examples of tube bends.

Type Bender	AB	AB	B	B	B	BC	B	BC	B	BC	C	BC	C
Tube od	⅛"	³⁄₁₆"	¼"	⁵⁄₁₆"	⅜"	⅜"	⁷⁄₁₆"	½"	½"	⅝"	⅝"	¾"	¾"
Standard Bend	⅜"	⁷⁄₁₆"	⁹⁄₁₆"	¹¹⁄₁₆"	¹¹⁄₁₆"	¹⁵⁄₁₆"	1⅜"	1½"	1¼"	2"	1½"	2½"	1¾"

Type Bender	C	B	C	C	C	C	C	C	C	C	C	C	C
Tube od	⅞"	1"	1"	1⅛"	1¼"	1⅜"	1⅜"	1½"	1½"	1¾"	2"	2½"	3"
Standard Bend	2"	3½"	3"	3½"	3¾"	5"	6"	5"	6"	7"	8"	10"	12"

A–Hand   B–Portable hand benders   C–Production bender

Figure 7-11   Minimum tube bend radii chart.

Figure 7-11 shows the minimum bend radii for tubing using hand benders and production benders. The mechanic should always consult the minimum bend radius chart before bending tubing because damage to the tubing could result from bends that are made to tight.

Tube flaring

The flaring tools (Fig. 7-12) used for aircraft tubing has male and female dies ground to produce a flare of 35 to 37 degrees. Under no

Figure 7-12   Single and double flaring tools.

circumstances is it permissible to use an automotive flaring tool, which produces a 45° flare.

Two kinds of flares are generally used in aircraft plumbing systems: single and double (Fig. 7.13).

In forming flares, cut the tube ends square, file them smooth, remove all burrs and sharp edges, and thoroughly clean the edges. Slip the fitting nut and sleeve on the tube before flaring it.

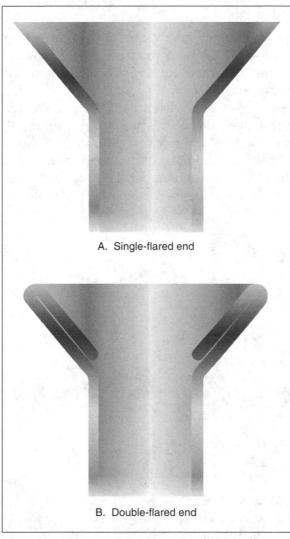

A. Single-flared end

B. Double-flared end

Figure 7-13   Single and double flares.

Assembling sleeve-type fittings

Sleeve-type end fittings for flexible hose are detachable and can be reused if they are determined to be serviceable. The inside diameter of the fitting is the same as the inside diameter of the hose to which it is attached. Common sleeve-type fittings are shown in Fig. 7-14.

Refer to manufacturer's instructions for detailed assembly procedures, as outlined in Fig. 7-15.

Proof-testing after assembly

All flexible hose must be proof-tested after assembly by plugging or capping one end of the hose and applying pressure to the inside of the hose assembly. The proof-test medium can be a liquid or a gas. For example, hydraulic, fuel, and oil lines are generally tested using hydraulic oil or water, whereas air or instrument lines are tested with dry, oil-free air or nitrogen. When testing with a liquid, all trapped air is bled from the assembly prior to tightening the cap or plug. Hose tests, using a gas, are conducted underwater. In all cases, follow the hose manufacturer's instructions for the proof-test pressure and fluid to be used when testing a specific hose assembly.

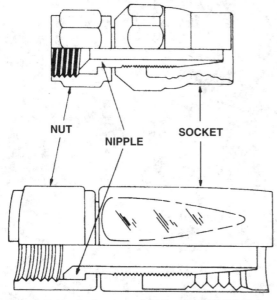

Figure 7-14   A sleeve end fitting for flexible hose.

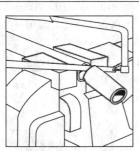

1. Place hose in vise and cut to desired length using fine tooth hacksaw or cut off wheel.

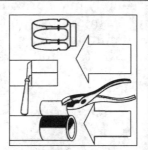

2. Locate length of hose to be cut off and slit cover with knife to wire braid. After slitting cover, twist off with pair of pliers.
(See note below.)

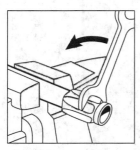

3. Place hose in vise and screw socket on hose counterclockwise.

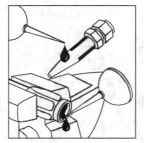

4. *Lubricate inside of hose and nipple threads liberally.

NOTE: Hose assemblies fabricated per MIL-H-8790 must have the exposed wire braid coated with a special sealant.

NOTE: Step 2 applies to high pressure hose only.

*CAUTION: Do not use any petroleum product with hose designed for synthetic fluids (SKYDROL and/or HYJET product). For a lubricant during assembly, use a vegetable soap liquid.

Disassemble in reverse order.

5. Screw nipple into socket using wrench on hex of nipple and leave .005" to .031" clearance between nipple hex and socket.

Figure 7-15 Assemble of MS fitting to flexible hose.

Place the hose assembly in a horizontal position and observe it for leakage while maintaining the test pressure. Proof-test pressures should be maintained for at least 30 seconds.

Figure 7-16 shows the test and burst pressures for flexible aircraft hose.

SINGLE WIRE BRAID FABRIC COVERED

MIL. Part No.	Tube Size o.d. (inches)	Hose Size i.d. (inches)	Hose Size o.d. (inches)	Recomm. Operating Pressure (PSI)	Min. Burst Pressure (PSI)	Max. Proof Pressure (PSI)	Min. Bend Radius (inches)
MIL-H-8794-3-L	3/16	1/8	.45	3,000	12,000	6,000	3.00
MIL-H-8794-4-L	1/4	3/16	.52	3,000	12,000	6,000	3.00
MIL-H-8794-5-L	5/16	1/4	.58	3,000	10,000	5,000	3.38
MIL-H-8794-6-L	3/8	5/16	.67	2,000	9,000	4,500	4.00
MIL-H-8794-8-L	1/2	13/32	.77	2,000	8,000	4,000	4.63
MIL-H-8794-10-L	5/8	1/2	.92	1,750	7,000	3,500	5.50
MIL-H-8794-12-L	3/4	5/8	1.08	1,750	6,000	3,000	6.50
MIL-H-8794-16-L	1	7/8	1.23	800	3,200	1,600	7.38
MIL-H-8794-20-L	1 1/4	1 1/8	1.50	600	2,500	1,250	9.00
MIL-H-8794-24-L	1 1/2	1 3/8	1.75	500	2,000	1,000	11.00
MIL-H-8794-32-L	2	1 13/16	2.22	350	1,400	700	13.25
MIL-H-8794-40-L	2 1/2	2 3/8	2.88	200	1,000	300	24.00
MIL-H-8794-48-L	3	3	3.56	200	800	300	33.00

Construction: Seamless synthetic rubber inner tube reinforced with one fiber braid, one braid of high tensile steel wire and covered with an oil resistant rubber impregnated fiber braid.

Identification: Hose is identified by specification number, size number, quarter year and year, hose manufacturer's identification.

Uses: Hose is approved for use in aircraft hydraulic, pneumatic, coolant, fuel, and oil systems.

Operating Temperatures:
Sizes 3 through 12: Minus 65°F to plus 250°F
Sizes 16 through 48: Minus 40°F to plus 275°F

Note: Maximum temperatures and pressures should not be used simultaneously.

MULTIPLE WIRE BRAID RUBBER COVERED

MIL. Part No.	Tube Size o.d. (inches)	Hose Size i.d. (inches)	Hose Size o.d. (inches)	Recomm. Operating Pressure (PSI)	Min. Burst Pressure (PSI)	Max. Proof Pressure (PSI)	Min. Bend Radius (inches)
MIL-H-8788- 4-L	1/4	7/32	.63	3,000	16,000	8,000	3.00
MIL-H-8788- 5-L	5/16	9/32	.70	3,000	14,000	7,000	3.38
MIL-H-8788- 6-L	3/8	11/32	.77	3,000	14,000	7,000	5.00
MIL-H-8788- 8-L	1/2	7/16	.86	3,000	14,000	7,500	5.75
MIL-H-8788-10-L	5/8	9/16	1.03	3,000	12,000	6,000	6.50
MIL-H-8788-12-L	3/4	11/16	1.22	3,000	12,000	6,000	7.75
MIL-H-8788-16-L	1	7/8	1.50	3,000	10,000	5,000	9.63

Construction: Seamless synthetic rubber inner tube reinforced with one fiber braid, two or more steel wire braids, and covered with synthetic rubber cover (for gas applications request perforated cover).

Identification: Hose is identified by specification number, size number, quarter year and year, hose manufacturer's identification.

Uses: High pressure hydraulic, pneumatic, coolant, fuel and oil.

Operating Temperatures: Minus 65°F to plus 200°F

Figure 7-16 Aircraft hose specifications.

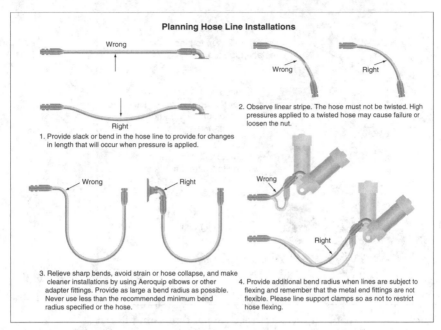

Figure 7-17 Installation of flexible hose assemblies.

Installing flexible hose assemblies

Figure 7-17 shows examples of flexible hose installation.

Installing Rigid Tubing

Never apply compound to the faces of the fitting or the flare because the compound will destroy the metal-to-metal contact between the fitting and flare, a contact that is necessary to create the seal. Be sure that the line assembly is properly aligned before tightening the fittings. Do not pull the installation into place with torque on the nut (Fig. 7-18).

Always tighten fittings to the correct torque value (Fig. 7-19) when installing a tube assembly. Overtightening a fitting might badly damage or completely cut off the tube flare, or it might ruin the sleeve or fitting nut. Failure to tighten sufficiently also can be serious; it might allow the line to blow out of the assembly or to leak under system pressure.

The use of torque wrenches and the prescribed torque values prevents overtightening or undertightening. If a tube-fitting assembly is tightened properly, it can be removed and retightened many times before reflaring is necessary.

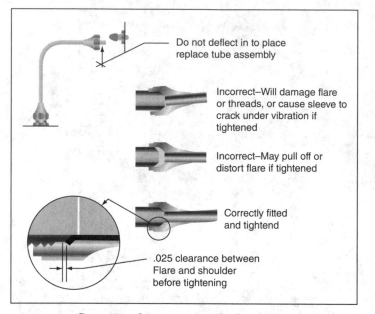

Do not deflect in to place
replace tube assembly

Incorrect–Will damage flare
or threads, or cause sleeve to
crack under vibration if
tightened

Incorrect–May pull off or
distort flare if tightened

Correctly fitted
and tightend

.025 clearance between
Flare and shoulder
before tightening

Figure 7-18   Correct and incorrect methods of tightening flared
tube fittings (*courtesy Aeroquip Corporation*).

Never select a path that does not require bends in the tubing. A
tube cannot be cut or flared accurately enough that it can be
installed without bending and still be free from mechanical strain.
Bends are also necessary to permit the tubing to expand or contract
under temperature changes and to absorb vibration. If the tube is
small (less than ¼ inch) and can be hand formed, casual bends can
be made to allow for this. If the tube must be machine formed,
definite bends must be made to avoid a straight assembly.

Start all bends a reasonable distance from the fittings because
the sleeves and nuts must be slipped back during the fabrication
of flares and during inspections. In all cases, the new tube assembly should be so formed prior to installation that it will not be
necessary to pull or deflect the assembly into alignment by means
of the coupling nuts.

Support clamps

Support clamps are used to secure the various lines to the airframe
or power-plant assemblies. Several types of support clamps are used
for this purpose, most commonly the rubber-cushioned and plain
clamps. The rubber-cushioned clamp is used to secure lines subject

TUBING O.D.	FITTING BOLT OR NUT SIZE	ALUMINUM ALLOY TUBING, BOLT FITTING OR NUT TORQUE INCH-LBS.	STEEL TUBING, BOLT FITTING OR NUT TORQUE INCH-LBS.	HOSE END FITTINGS AND HOSE ASSEMBLIES MS 28740 OR EQUIVALENT END FITTING		MINIMUM BEND RADII (INCHES)	
				MINIMUM	MAXIMUM	ALUM. ALLOY 1100–H14 5052–O	STEEL
⅛ –2	–2	20–30				⅜	
3⁄16 –3	–3	30–40	90–100	70	120	7⁄16	21⁄32
¼ –4	–4	40–65	135–150	100	250	9⁄16	⅞
5⁄16 –5	–5	60–85	180–200	210	420	¾	1⅛
⅜ –6	–6	75–125	270–300	300	480	15⁄16	1 1⁄16
½ 8	–8	150–250	450–500	500	850	1¼	1¾
⅝ –10	–10	200–350	650–700	700	1150	1½	2 3⁄16
¾ –12	–12	300–500	900–1000			1¾	2⅝
⅞ –14	–14	500–600	1000–1100				
1 –16	–16	500–700	1200–1400			3	3½
1¼ –20	–20	600–900	1200–1400			3¾	4⅜
1½ –24	–24	600–900	1500–1800			5	5¼
1¾ –28	–28	850–1050				7	6⅛
2 –32	–32	950–1150				8	7

Figure 7-19 Torque values for tightening flared tube fittings.

to vibration; the cushioning prevents chafing of the tubing. The plain clamp is used to secure lines in areas not subject to vibration.

A Teflon-cushioned clamp is used in areas where the deteriorating effect of Skydrol 500, hydraulic fluid (MIL-0-5606), or fuel is expected. However, because Teflon is less resilient, it does not provide as good of a vibration-damping effect as other cushion materials.

Use bonded clamps to secure metal hydraulic, fuel, and oil lines in place. Unbonded clamps should be used only to secure wiring. Remove any paint or anodizing from the portion of the tube at the bonding clamp location. All plumbing lines must be secured at specified intervals. The maximum distance between supports for rigid tubing is shown in Fig. 7-20.

Rigid tubing inspection and repair

Minor dents and scratches in tubing may be repaired. Scratches or nicks not deeper than 10 percent of the wall thickness in aluminum alloy tubing, which are not in the heel of a bend, may be repaired by burnishing with hand tools. The damage limits for hard, thinwalled corrosion-resistant steel and titanium tubing are considerably less than for aluminum tubing, and might depend on the aircraft manufacturer. Consult the aircraft maintenance manual for damage limits. Replace lines with severe die marks, seams, or splits in the tube. Any crack or deformity in a flare is unacceptable and is cause for rejection. A dent of less than 20 percent of the tube diameter is not objectionable, unless it is in the heel of a bend. A severely damaged line should be replaced.

TUBE OD	DISTANCE BETWEEN SUPPORTS (IN.)	
(IN.)	ALUMINUM ALLOY	STEEL
1/8	9½	11½
3/16	12	14
1/4	13½	16
5/16	15	18
3/8	16½	20
1/2	19	23
5/8	22	25½
3/4	24	27½
1	26½	30

Figure 7-20 Maximum distance between supports for fluid lines.

However, the line may be repaired by cutting out the damaged section and inserting a tube section of the same size and material. Flare both ends of the undamaged and replacement tube sections and make the connection by using standard unions, sleeves, and tube nuts. Aluminum 6061-T6, corrosion resistant steel 304-1/8h, and Titanium 3AL-2.5V tubing can be repaired by swaged fittings. If the damaged portion is short enough, omit the insert tube and repair by using one repair union as shown in Fig. 7-21. When repairing a damaged line, be very careful to remove all chips and burrs. Any open line that is to be left unattended for some time

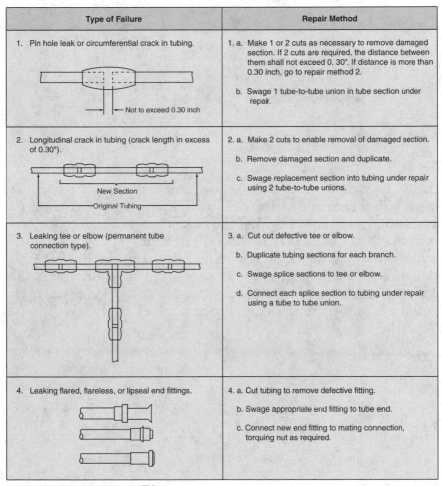

Type of Failure	Repair Method
1. Pin hole leak or circumferential crack in tubing. Not to exceed 0.30 inch	1. a. Make 1 or 2 cuts as necessary to remove damaged section. If 2 cuts are required, the distance between them shall not exceed 0. 30". If distance is more than 0.30 inch, go to repair method 2. b. Swage 1 tube-to-tube union in tube section under repair.
2. Longitudinal crack in tubing (crack length in excess of 0.30"). New Section Original Tubing	2. a. Make 2 cuts to enable removal of damaged section. b. Remove damaged section and duplicate. c. Swage replacement section into tubing under repair using 2 tube-to-tube unions.
3. Leaking tee or elbow (permanent tube connection type).	3. a. Cut out defective tee or elbow. b. Duplicate tubing sections for each branch. c. Swage splice sections to tee or elbow. d. Connect each splice section to tubing under repair using a tube to tube union.
4. Leaking flared, flareless, or lipseal end fittings.	4. a. Cut tubing to remove defective fitting. b. Swage appropriate end fitting to tube end. c. Connect new end fitting to mating connection, torquing nut as required.

Figure 7-21 Permaswage™ repair.

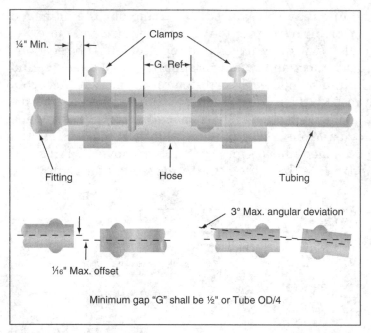

Figure 7-22   Repair of low-pressure rigid tube with a flexible hose.

should be sealed, using metal, wood, rubber, or plastic plugs or caps. When repairing a low-pressure line using a flexible fluid connection assembly, position the hose clamps carefully to prevent overhang of the clamp bands or chafing of the tightening screws on adjacent parts. If chafing can occur, the hose clamps should be repositioned on the hose. Figure 7-22 illustrates the design of a flexible fluid connection assembly and gives the maximum allowable angular and dimensional offset.

8

Control Cables

Three mechanical control systems commonly used are cable, push-pull (Fig. 8-1), and torque tube. Many aircraft incorporate control systems that are combinations of all three.

Cables are the most widely used linkage in primary flight control systems. Cable linkage is also used in engine controls, emergency extension systems for the landing gear, and other systems throughout the aircraft.

Cable Assembly

The conventional cable assembly consists of flexible cable (Fig. 8-2) terminals (end fittings) for attaching to other units, and turnbuckles. Cable tension must be adjusted frequently because of stretching and temperature changes. Aircraft-control cables are fabricated from carbon steel or stainless steel.

Fabricating a cable assembly

Terminals for aircraft-control cables are normally fabricated using three different processes:

- Swaging, as used in all modern aircraft.
- Nicopress process.
- Handwoven splice terminal.

Handwoven splices are used in many older aircraft; however, this time-consuming process is considered unnecessary with the

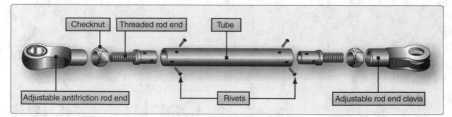

Figure 8-1   Push-pull tube assembly.

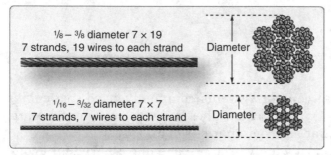

Figure 8-2   7 × 19 extra flexibility and 7 × 7 medium flexibility aircraft cables.

availability of mechanically fabricated splices. Various swage terminal fittings are shown in Fig. 8-3.

Swaging

Swage terminals, manufactured in accordance with Air Force/Navy Aeronautical Standard Specifications, are suitable for use in civil aircraft up to and including maximum cable loads. When swaging tools are used, it is important that all the manufacturers' instructions, including "go-no-go" dimensions (Fig. 8-4), are followed in detail to avoid defective and inferior swaging. Observance of all instructions should result in a terminal developing the full-rated strength of the cable.

Nicopress process

A patented process using copper sleeves may be used up to the full rated strength of the cable when the cable is looped around a thimble as shown in Fig. 8-5. This process may also be used

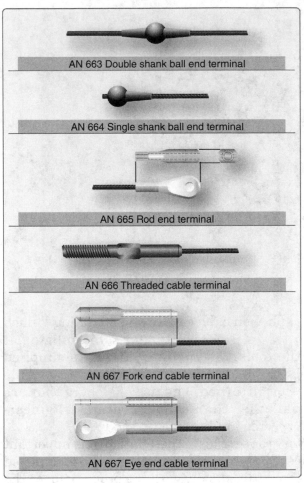

Figure 8-3 Various types of swaged terminal fittings.

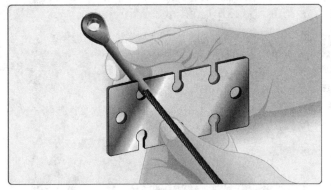

Figure 8-4 A typical gauge for checking swaged terminals.

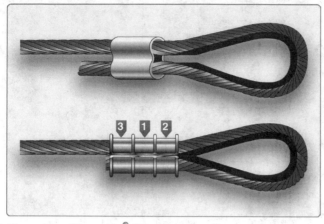

Figure 8-5 Nicopress® thimble-eye splice.

in place of the 5-tuck splice on cables up to and including ³/₈-inch diameter. Whenever this process is used for cable splicing, it is imperative that the tools, instructions, and data supplied by Nicopress® be followed exactly to ensure the desired cable function and strength is attained. The use of sleeves that are fabricated of material other than copper requires engineering approval for the specific application by the FAA.

To make a satisfactory copper sleeve installation, it is important that the amount of sleeve pressure be kept uniform. The completed sleeves should be checked periodically with the proper gauge. The gauge should be held so that it contacts the major axis of the sleeve. The compressed portion at the center of the sleeve should enter the gauge opening with very little clearance, as shown in Fig. 8-6. If it does not, the tool must be adjusted accordingly.

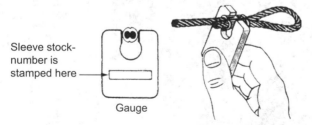

Sleeve stock-
number is
stamped here

Gauge

Figure 8-6 Typical go-no-go gauge for nicopress terminals.

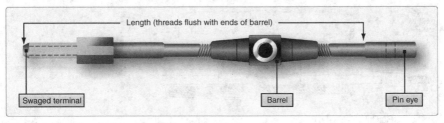

Figure 8-7 A typical turnbuckle assembly.

Turnbuckles

A turnbuckle assembly is a mechanical screw device that consists of two threaded terminals and a threaded barrel. Figure 8-7 illustrates a typical turnbuckle assembly.

Turnbuckles are fitted in the cable assembly for the purpose of making minor adjustments in cable length and to adjust cable tension. One of the terminals has right-handed threads and the other has left-handed threads. The barrel has matching right- and left-handed internal threads. The end of the barrel with the left-handed threads can usually be identified by a groove or knurl around that end of the barrel.

When installing a turnbuckle in a control system, it is necessary to screw both of the terminals an equal number of turns into the barrel. It is also essential that all turnbuckle terminals be screwed into the barrel until not more than three threads are exposed on either side of the turnbuckle barrel. After a turnbuckle is properly adjusted, it must be safetied.

Safety methods for turnbuckles

After a turnbuckle has been properly adjusted, it must be safetied. There are several methods of safetying turnbuckles; however, only two methods (Figs. 8-8 and 8-9) are covered in this chapter. The clip-locking method (Fig. 8-8) is used only on modern aircraft. Older aircraft still use turnbuckles that require the wire-wrapping method.

Double-wrap method

Of the methods using safety wire for safetying turnbuckles, the double-wrap method is preferred, although the single-wrap

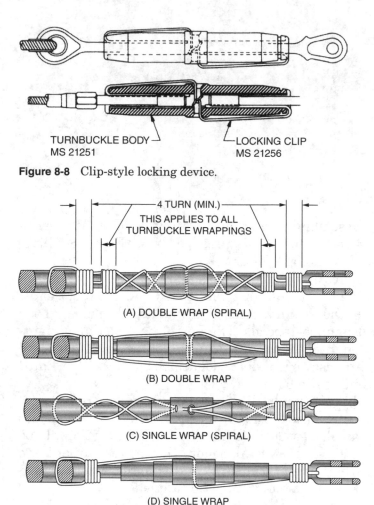

Figure 8-8 Clip-style locking device.

TURNBUCKLE BODY
MS 21251

LOCKING CLIP
MS 21256

4 TURN (MIN.)
THIS APPLIES TO ALL
TURNBUCKLE WRAPPINGS

(A) DOUBLE WRAP (SPIRAL)

(B) DOUBLE WRAP

(C) SINGLE WRAP (SPIRAL)

(D) SINGLE WRAP

Figure 8-9 Turnbuckle safety wiring methods.

method is satisfactory. The method of double-wrap safetying is shown in Fig. 8-9. Two separate lengths of the proper wire, as shown in Fig. 8-10, are used. One end of the wire is run through the hole in the barrel of the turnbuckle. The ends of the wire are bent toward opposite ends of the turnbuckle.

Then the second length of the wire is passed into the hole in the barrel with the ends bent along the barrel on the side opposite of the first. Then the wires at the end of the turnbuckle are passed in opposite directions through the holes in the turnbuckle eyes or between the jaws of the turnbuckle fork, as applicable.

Cable Size (in.)	Type of Wrap	Diameter of Safety Wire	Material (Annealed Condition)
1/16	Single	0.020	Copper, brass.[1]
3/32	Single	0.040	Copper, brass.[1]
1/8	Single	0.040	Stainless steel, Monel and "K" Monel.
1/8	Double	0.040	Copper, brass.[1]
1/8	Single	0.057 min	Copper, brass.[1]
5/32 and greater	Double	0.040	Stainless steel, Monel and "K" Monel.[1]
5/32 and greater	Single	0.057 min	Stainless steel, Monel or "K" Monel.[1]
5/32 and greater	Double	0.051[2]	Copper, brass.

[1] Galvanized or tinned steel, or soft iron wires are also acceptable.
[2] The safety wire holes in 5/32-inch diameter and larger turnbuckle terminals for swaging may be drilled sufficiently to accommodate the double 0.051-inch diameter copper or brass wires when used.

Figure 8-10 Guide for selecting turnbuckle safety wire.

The laid wires are bent in place before cutting off the wrapped wire. The remaining length of safety wire is wrapped at least four turns around the shank, and cut off. The procedure is repeated at the opposite end of the turnbuckle.

When a swaged terminal is being safetied, the ends of both wires are passed, if possible, through the hole provided in the terminal for this purpose and both ends are wrapped around the shank, as described previously.

If the hole is not large enough to allow passage of both wires, the wire should be passed through the hole and looped over the free end of the other wire, and then both ends are wrapped around the shank, as described.

Cable Tension Adjustment

Control cable tension should be carefully adjusted, in accordance with the air-frame manufacturer's recommendations. On large aircraft, the temperature of the immediate area should be taken into consideration when using a tensionmeter (Fig. 8-11). For long cable sections, the average of two or three temperature readings should be made for extreme surface temperature

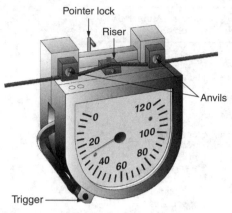

Figure 8-11 Typical cable tensionmeter.

variations that might be encountered if the aircraft is operated primarily in unusual geographic or climatic conditions, such as arctic, arid, or tropical locations. Figure 8-12 shows a typical cable rigging chart.

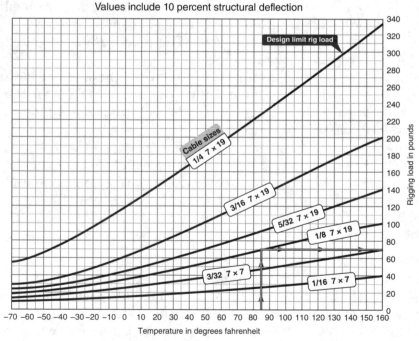

Figure 8-12 Typical cable rigging chart.

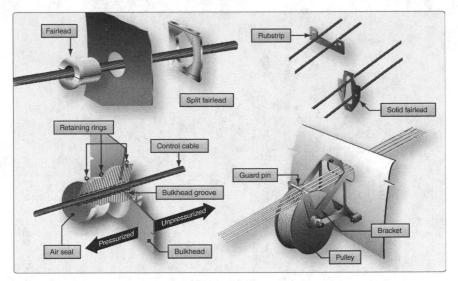

Figure 8-13 Various types of control cable guides.

Cable guides

Pulleys are used to guide cables and also to change the direction of cable movement. Pulley bearings are sealed and need no lubrication other than the lubrication done at the factory. Brackets fastened to the structure of the aircraft support the pulleys. Cables passing over pulleys are kept in place by guards. The guards are close fitting to prevent jamming or to prevent the cables from slipping off when they slacken due to temperature variations. Fairleads may be made from a nonmetallic material, such as phenolic, or a metallic material, such as soft aluminum. The fairlead completely encircles the cable where it passes through holes in bulkheads or other metal parts. Fairleads are used to guide cables in a straight line through or between structural members of the aircraft. Fairleads should never deflect the alignment of a cable more than 3 degree from a straight line. Pressure seals are installed where cables (or rods) move through pressure bulkheads. The seal grips tightly enough to prevent excess air pressure loss but not enough to hinder movement of the cable. Figure 8-13 shows the various types of cable guides.

9

Electrical Wiring and Installation

Material Selection

Aircraft service imposes severe environmental conditions on electrical wire. To ensure satisfactory service, the wire should be inspected at regular intervals for abrasions, defective insulation, condition of terminal posts, and corrosion under or around swaged terminals.

For the purpose of this section, a wire is described as a single, solid conductor, or as a stranded conductor covered with an insulating material (Fig. 9-1).

The term *cable*, as used in aircraft electrical installations, includes:

1. Two or more separately insulated conductors in the same jacket (multiconductor cable).

2. Two or more separately insulated conductors twisted together (twisted pair).

3. One or more insulated conductors, covered with a metallic braided shield (shielded cable).

4. A single insulated center conductor with a metallic braided outer conductor (radio-frequency cable). The concentricity of the center conductor and the outer conductor is carefully controlled during manufacturing to ensure that they are coaxial.

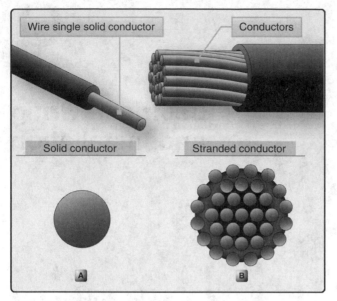

Wire single solid conductor

Conductors

Solid conductor

Stranded conductor

A

B

Figure 9-1 Aircraft electrical wire.

The term "wire harness" is used when an array of insulated conductors are bound together by lacing cord, metal bands, or other binding in an arrangement suitable for use only in specific equipment for which the harness was designed; it may include terminations. Wire harnesses are extensively used in aircraft to connect all the electrical components.

Wire size

Wire is manufactured in sizes according to a standard known as the *AWG (American wire gauge)*. As shown in Fig. 9-2, the wire diameters become smaller as the gauge numbers become larger. See the appendix for a table of wire gauges.

To use the wire gauge, the wire to be measured is inserted in the smallest slot that will accommodate the bare wire. The gauge number corresponding to that slot indicates the wire size. The slot has parallel sides and should not be confused with the semicircular opening at the end of the slot. The opening simply permits the free movement of the wire all the way through the slot.

Gauge numbers are useful in comparing the diameter of wires, but not all types of wire or cable can be accurately measured with

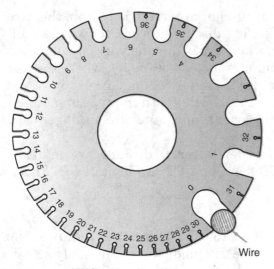

Wire

Figure 9-2 AWG wire gauge.

a gauge. Large wires are usually stranded to increase their flexibility. In such cases, the total area can be determined by multiplying the area of one strand (usually computed in circular mils when the diameter or gauge number is known) by the number of strands in the wire or cable.

Factors that affect the selection of wire size.   Tables and procedures are available for selecting correct wire sizes. For purposes of this manual, it is assumed that wire sizes were specified by the manufacturer of the aircraft or equipment.

Factors that affect the selection of conductor material.   Although silver is the best conductor, high cost limits its use to special circuits where a substance with high conductivity is needed.

The two most generally used conductors are copper and aluminum. Each has characteristics that make its use advantageous under certain circumstances. Also, each has certain disadvantages.

Copper has a higher conductivity; it is more ductile (can be drawn), has relatively high tensile strength, and can be easily soldered. It is more expensive and heavier than aluminum.

Although aluminum has only about 60 percent of the conductivity of copper, it is used extensively. Its lightness makes possible long spans, and its relatively large diameter for a given conductivity reduces corona, which is the discharge of electricity from the wire when it has a high potential. The discharge is greater when small-diameter wire is used than when large-diameter wire is used. Some bus bars are made of aluminum instead of copper, where there is a greater radiating surface for the same conductance.

Conductor insulation material varies with the type of installation. Such insulation as rubber, silk, and paper are no longer used extensively in aircraft systems.

Insulation materials for new aircraft designs are made of Tefzel®, Teflon®/Kapton®/Teflon®, and PTFE/Polyimide/PTFE.

Stripping insulation

Attaching the wire to connectors or terminals requires the removal of insulation to expose the conductors, commonly known as *stripping*. When stripping the wire, remove no more insulation than is necessary. Stripping can be accomplished in many ways; however, the following basic principles should be followed:

- Be sure that all cutting tools used for stripping are sharp.
- When using special wire stripping tools, adjust the tool to avoid nicking, cutting, or otherwise damaging the strands. A light-duty hand-operated wire stripper is shown in Fig. 9-3.
- Automatic stripping tools should be carefully adjusted; the manufacturer's instructions should be followed to avoid nicking, cutting, or otherwise damaging strands. This is especially important for aluminum wires and for copper wires smaller than No. 10. Smaller wires have larger numbers.

Terminals

Terminals are attached to the ends of electric wires to facilitate connection of the wires to terminal strips or items of equipment. Terminals specifically designed for use with the standard sizes of aircraft wire are available through normal supply channels. A haphazard choice of commercial terminals can contribute to overheated joints, vibration failures, and corrosion difficulties.

For most applications, soldered terminals have been replaced by solderless terminals. The solder process has disadvantages that have been overcome by use of the solderless terminals.

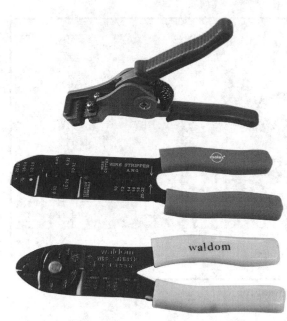

Figure 9-3 Wire strippers.

The terminal manufacturer will normally provide a special crimping or swaging tool for joining the solderless terminal to the electric wire. Aluminum wire presents special difficulty in that each individual strand is insulated by an oxide coating. This oxide coating must be broken down in the crimping process and some method used to prevent its reforming. In all cases, terminal manufacturer's instructions should be followed when installing solderless terminals.

Copper wires are terminated with solderless, preinsulated, straight copper terminal lugs. The insulation is part of the terminal lug and extends beyond its barrel so that it will cover a portion of the wire insulation, making the use of an insulation sleeve unnecessary (Fig. 9-4).

In addition, preinsulated terminal lugs contain an insulation grip (a metal reinforcing sleeve) beneath the insulation for extra gripping strength on the wire insulation. Preinsulated terminals accommodate more than one size of wire; the insulation is usually color-coded to identify the wire sizes that can be terminated with each of the terminal lug sizes.

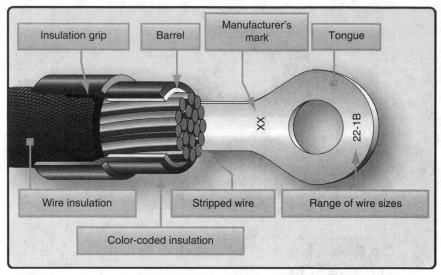

Figure 9-4 Preinsulated terminal lug.

Some types of uninsulated terminal lugs are insulated after assembly to a wire by means of pieces of transparent flexible tubing called sleeves. The sleeve provides electrical and mechanical protection at the connection. When the size of the sleeving used is such that it will fit tightly over the terminal lug, the sleeving need not be tied; otherwise, it should be tied with lacing cord, as illustrated in Fig. 9-5.

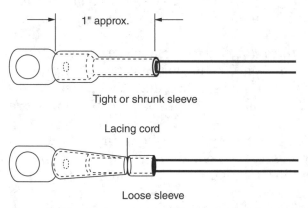

Figure 9-5 Insulating a terminal lug with a transparent, flexible tubing sleeve.

Aluminum wire terminals

The use of aluminum wire in aircraft systems is increasing because of its weight advantage over copper. However, bending aluminum will cause "work hardening" of the metal, making it brittle. This results in failure or breakage of strands much sooner than in a similar case with copper wire. Aluminum also forms a high-resistant oxide film immediately upon exposure to air. To compensate for these disadvantages, it is important to use the most reliable installation procedures.

Only aluminum terminal lugs are used to terminate aluminum wires. All aluminum terminals incorporate an inspection hole (Fig. 9-6), which permits checking the depth of wire insertion. The barrel of aluminum terminal lugs is filled with a petrolatum-zinc dust compound. This compound removes the oxide film from the aluminum by a grinding process during the crimping operation. The compound will also minimize later oxidation of the completed

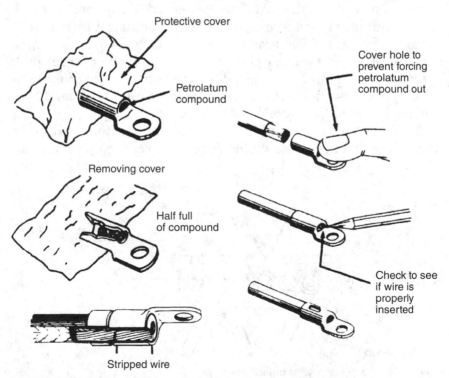

Figure 9-6 Inserting aluminum wire into aluminum terminal lugs.

connection by excluding moisture and air. The compound is retained inside the terminal lug barrel by a plastic or foil seal at the end of the barrel.

Connecting terminal lugs to terminal blocks

Terminal lugs should be installed on terminal blocks so that they are locked against movement in the direction of loosening (Fig. 9-7).

Terminal blocks are normally supplied with studs secured in place by a plain washer, an external tooth lockwasher, and a nut. In connecting terminals, it is recommended to place copper terminal lugs directly on top of the nut, followed with a plain washer and elastic stop nut, or with a plain washer, split steel lockwasher, and plain nut.

Aluminum terminal lugs should be placed over a plated brass plain washer, followed with another plated brass plain washer, split steel lockwasher, and plain nut or elastic stop nut. The plated brass washer should have a diameter equal to the tongue width of the aluminum terminal lug. The manufacturer's instructions should be consulted for recommended dimensions of these plated brass washers. No washer should be placed in the current path between two aluminum terminal lugs or between two copper terminal lugs. Also, no lockwasher should be placed against the tongue or pad of the aluminum terminal.

To join a copper terminal lug to an aluminum terminal lug, a plated brass plain washer should be placed over the nut that holds

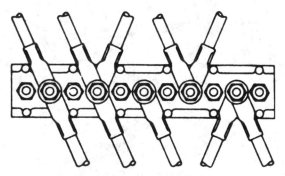

Figure 9-7 Connecting terminals to a terminal block.

the stud in place, followed with the aluminum terminal lug, a plated brass plain washer, the copper terminal lug, plain washer, split steel lockwasher, and a plain nut or a self-locking, all-metal nut. As a general rule, a torque wrench should be used to tighten nuts to ensure sufficient contact pressure. Manufacturer's instructions provide installation torques for all types of terminals.

Wiring identification

The proper identification of electrical wires and cables with their circuits and voltages is necessary to provide safety of operation, safety to maintenance personnel, and ease of maintenance. All wire used on aircraft must have its type identification imprinted along its length. It is common practice to follow this part number with the five digit/letter Commercial and Government Entity (CAGE) code identifying the wire manufacturer. You can identify the performance capabilities of existing installed wire you need to replace, and avoid the inadvertent use of a lower performance and unsuitable replacement wire.

Placement of identification markings

Identification markings should be placed at each end of the wire and at 15-inch maximum intervals along the length of the wire. Wires less than 3 inches in length need not be identified. Wires 3 to 7 inches in length should be identified approximately at the center. Added identification marker sleeves should be located so that ties, clamps, or supporting devices need not be removed to read the identification. The wire identification code must be printed to read horizontally (from left to right) or vertically (from top to bottom). The two methods of marking wire or cable are as follows:

1. Direct marking is accomplished by printing the cable's outer covering as shown in Fig. 9-8.

2. Indirect marking is accomplished by printing a heat-shrinkable sleeve and installing the printed sleeve on the wire or cables outer covering. Indirectly-marked wire or cable should be identified with printed sleeves at each end and at intervals not longer than 6 feet, see Fig. 9-9.

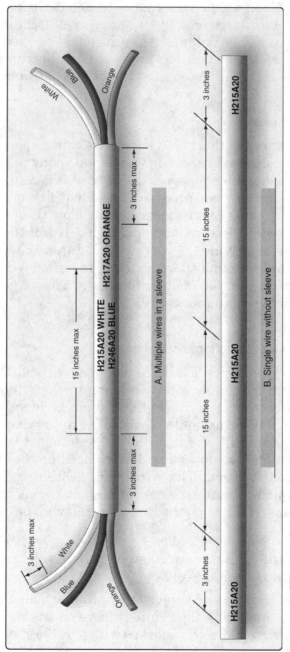

Figure 9-8 Wire markings for single wire without sleeve.

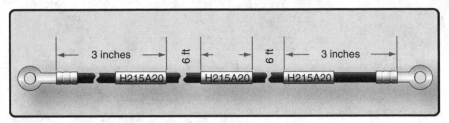

Figure 9-9 Spacing of printed identification marks.

Wire Groups and Bundles

Grouping or bundling certain wires, such as electrically unprotected power wiring and wiring going to duplicate vital equipment, should be avoided.

Wire bundles should generally contain fewer than 75 wires, or 1½ inch to 2 inch in diameter where practicable. When several wires are grouped at junction boxes, terminal blocks, panels, and the like, the identity of the group within a bundle (Fig. 9-10) can be retained.

The flexible nylon cable tie (Fig. 9-11) has almost completely replaced cord for lacing or tying wire bundles. Nylon cable ties are available in various lengths and are self-locking for a permanent, neat installation.

Figure 9-10 Groups and bundle ties.

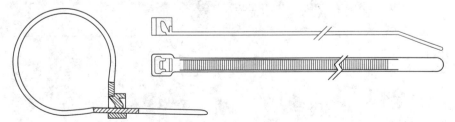

Figure 9-11 Flexible nylon cable ties have almost completely replaced cord for lacing or tying cable bundles.

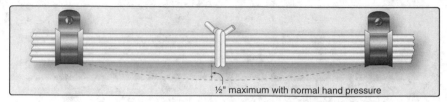

Figure 9-12   Maximum recommended slack in wire bundles between supports.

Single wires or wire bundles should not be installed with excessive slack. Slack between supports should normally not exceed a maximum of ½ inch deflection with normal hand force (Fig. 9-12).

Spliced connections in wire bundles

Splicing is permitted on wiring as long as it does not affect the reliability and the electromechanical characteristics of the wiring. Splicing of electrical wire should be kept to a minimum and avoided entirely in locations subject to extreme vibrations. Many types of aircraft splice connector are available for use when splicing individual wires. Use of a self-insulated splice connector is preferred; however, a noninsulated splice connector may be used provided the splice is covered with plastic sleeving that is secured at both ends.

- There should be no more than one splice in any one wire segment between any two connectors or other disconnect points. Exceptions include when attaching to the spare pigtail lead of a potted connector, when splicing multiple wires to a single wire, when adjusting wire size to fit connector contact crimp barrel size, and when required to make an approved repair.

- Splices in bundles must be staggered to minimize any increase in the size of the bundle, preventing the bundle from fitting into its designated space or causing congestion that adversely affects maintenance as shown in Fig. 9-13.

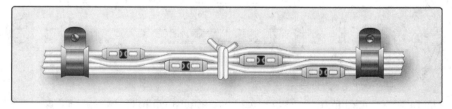

Figure 9-13   Staggered placement of splices in a wire bundle.

- Splices should not be used within 12 inches of a termination device, except when attaching to the pigtail spare lead of a potted termination device, to splice multiple wires to a single wire, or to adjust the wire sizes so that they are compatible with the contact crimp barrel sizes.

Bend Radii

Bends in wire groups or bundles should not be less than 10 times the outside diameter of the wire group or bundle. However, at terminal strips, where wire is suitably supported at each end of the bend, a minimum radius of three times the outside diameter of the wire, or wire bundle, is normally acceptable. There are, of course, exceptions to these guidelines in the case of certain types of cable; for example, coaxial cable should never be bent to a smaller radius than six times the outside diameter.

Routing and installations

All wiring should be installed so that it is mechanically and electrically sound and neat in appearance. Whenever practicable, wires and bundles should be routed parallel with, or at right angles to, the stringers or ribs of the area involved. An exception to this general rule is coaxial cable, which is routed as directly as possible.

The wiring must be adequately supported throughout its length. A sufficient number of supports must be provided to prevent undue vibration of the unsupported lengths. Wire clamps should be spaced at intervals not exceeding 24 inches.

When wiring must be routed parallel to combustible fluid or oxygen lines for short distances, as much fixed separation as possible should be maintained. The wires should be on a level with, or above, the plumbing lines. Clamps should be spaced so that if a wire is broken at a clamp, it will not contact the line. Where a 6 inch separation is not possible, both the wire bundle and the plumbing line can be clamped to the same structure to prevent any relative motion. If the separation is less than 2 inch, but more than ½ inch, a polyethylene sleeve can be used over the wire bundle to give further protection. Also, two cable clamps back-to-back, as shown in Fig. 9-14, can be used to maintain a rigid separation only, and not for support of the bundle. No wire should be routed so that it is located nearer than ½ inch to a plumbing line.

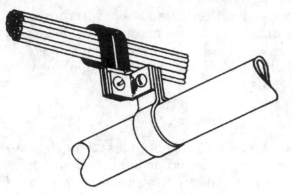

Figure 9-14 Method of separating wires from plumbing lines.

Neither should a wire or wire bundle be supported from a plumbing line that carries flammable fluids or oxygen.

Wiring should be routed to maintain a minimum clearance of at least 3 inch from control cables. If this cannot be accomplished, mechanical guards should be installed to prevent contact between the wiring and control cables.

Cable clamps should be installed with regard to the proper angle, as shown in Fig. 9-15. The mounting screw should be above the wire bundle. It is also desirable that the back of the cable clamp rest against a structural member where practicable.

Care should be taken that wires are not pinched in cable clamps. Where possible, the cables should be mounted directly to structural members, as shown in Figs. 9-17 and 9-18. Clamps can be used with rubber cushions to secure wire bundles to tubular structures. Such clamps must fit tightly, but should not be deformed when locked in place.

Protection against chafing

Wires and wire groups should be protected against chafing or abrasion in those locations where contact with sharp surfaces or other wires would damage the insulation. Damage to the insulation can cause short circuits, malfunction, or inadvertent operation of equipment. Cable clamps should be used to support wire bundles at each hole through a bulkhead (Fig. 9-18). If wires come closer than ¼ inch to the edge of the hole, a suitable grommet should be used in the hole, as shown in Fig. 9-18.

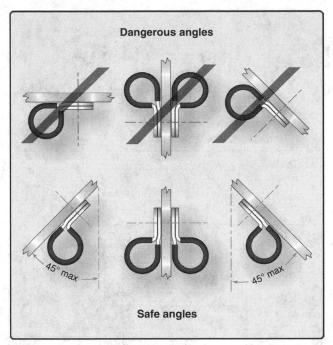

Figure 9-15 Proper and improper angles for installation of cable clamps.

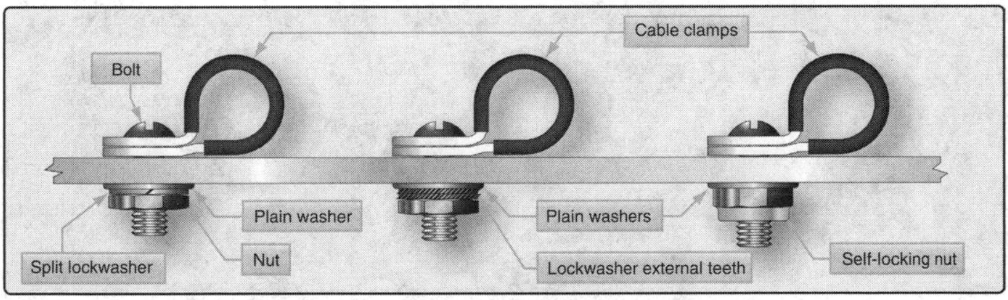

Figure 9-16 Various methods of mounting cable clamps.

Bonding and Grounding

Bonding is the electrical connecting of two or more conducting objects not otherwise adequately connected. Grounding is the electrical connecting of a conducting object to the primary structure for a return path for current. Primary structure is

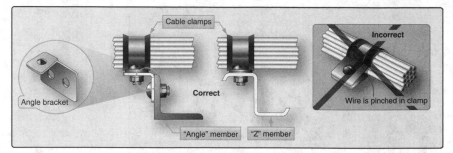

Figure 9-17 Mounting cable clamp to structure.

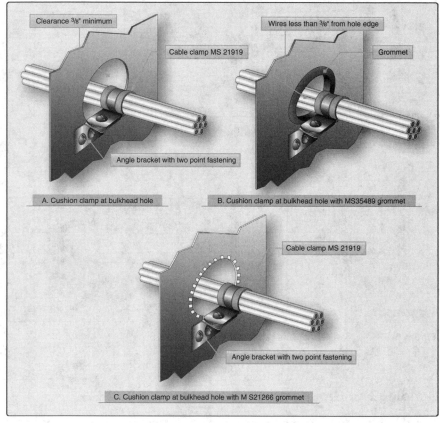

Figure 9-18 Cable clamps at bulkhead opening.

the main frame, fuselage, or wing structure of the aircraft, commonly referred to as *ground*. Bonding and grounding connections are made in aircraft electrical systems to:

- Protect aircraft and personnel against hazards from lightning discharge.
- Provide current return paths.
- Prevent development of radio frequency potentials.
- Protect personnel from shock hazards.
- Provide stability of radio transmission and reception.
- Prevent accumulation of static charge.

Bonding jumpers should be made as short as practicable, and installed in such manner that the resistance of each connection does not exceed 0.003 Ω. The jumper must not interfere with the operation of movable aircraft elements, such as surface controls, nor should the normal movement of these elements result in damage to the bonding jumper.

To ensure a low-resistance connection, nonconducting finishes, such as paint and anodizing films, should be removed from the attachment surface to be contacted by the bonding terminal. Electric wiring should not be grounded directly to magnesium parts.

Electrolytic action can rapidly corrode a bonding connection if suitable precautions are not taken. Aluminum alloy jumpers are recommended for most cases; however, copper jumpers should be used to bond together parts made of stainless steel, cadmium-plated steel, copper, brass, or bronze. Where contact between dissimilar metals cannot be avoided, the choice of jumper and hardware should be such that corrosion is minimized, and the part likely to corrode would be the jumper or associated hardware. Figure 9-19 shows the proper hardware combination for making a bond connection. At locations where finishes are removed, a protective finish should be applied to the completed connection to prevent subsequent corrosion.

The use of solder to attach bonding jumpers should be avoided. Tubular members should be bonded by means of clamps to which the jumper is attached. Proper choice of clamp material will minimize the probability of corrosion.

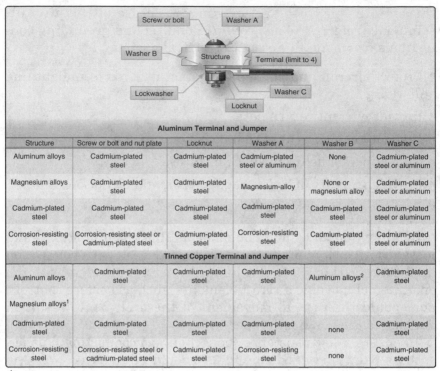

Figure 9-19   Bolt and nut bonding or grounding to flat surface.

AN/MS Connectors

Connectors (plugs and receptacles) facilitate maintenance when frequent disconnection is required. There is a multitude of types of connectors. The connector types that use crimped contacts are generally used on aircraft. Some of the more common types are the round cannon type, the rectangular, and the module blocks. Environmentally resistant connectors should be used in applications subject to fluids, vibration, heat, mechanical shock, and/or corrosive elements. Connectors must be identified by an original identification number derived from MIL Specification (MS) or OEM specification. Figure 9-20 provides information about MS style connectors.

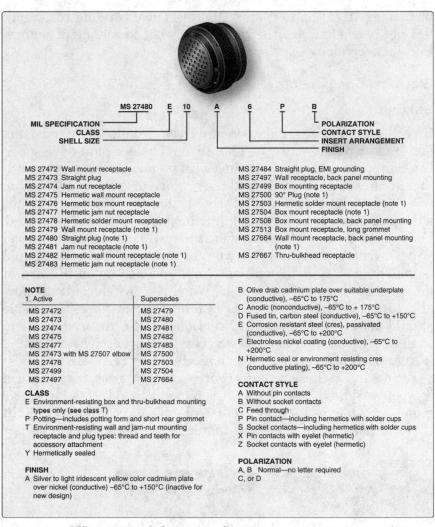

Figure 9-20 MS-connector information sheet.

Wire inspection

Aircraft service imposes severe environmental condition on electrical wire. To ensure satisfactory service, inspect wire annually for abrasions, defective insulation, condition of terminations, and potential corrosion. Grounding connections for power, distribution equipment, and electromagnetic shielding must be given particular attention to ensure that electrical bonding resistance has not been significantly increased by the loosening of connections or corrosion.

Electrical Components

Switches

Switches are devices that open and close circuits. They consist of one or more pair of contacts. The current in the circuit flows when the contacts are closed. Switches with momentary contacts actuate the circuit temporarily, and they return to the normal position with an internal spring when the switch is released. Switches with continuous contacts remain in position when activated. Hazardous errors in switch operation can be avoided by logical and consistent installation. Two-position on/off switches should be mounted so that the on position is reached by an upward or forward movement of the toggle. When the switch controls movable aircraft elements, such as landing gear or flaps, the toggle should move in the same direction as the desired motion. Inadvertent operation of a switch can be prevented by mounting a suitable guard over the switch as shown in Fig. 9-21.

Toggle and rocker switches control most of aircraft's electrical components. Rotary switches are commonly found on radio control panels and micro-switches are used to detect position or to limit travel of moving parts, such as landing gear, flaps, spoilers, etc.

Figure 9-21 Switch with guard.

Relays and solenoids

Relays are used to control the flow of large currents using a small current. A low-power DC circuit is used to activate the relay and control the flow of large AC currents. They also switch motors and other electrical equipment on and off and to protect them from overheating. A solenoid is a special type of relay that has a moving core. The electromagnet core in a relay is fixed. Solenoids are mostly used as mechanical actuators but can also be used for switching large currents. Relays are only used to switch currents.

Fuses

A fuse is placed in series with the voltage source and all current must flow through it. The fuse consists of a strip of metal that is enclosed in a glass or plastic housing. The metal strip has a low melting point and is usually made of lead, tin, or copper. When the current exceeds the capacity of the fuse the metal strip heats up and breaks. As a result of this, the flow of current in the circuit stops. There are two basic types of fuses: fast acting and slow blow. The fast-acting type opens very quickly when their particular current rating is exceeded. This is important for electric devices that can quickly be destroyed when too much current flows through them for even a very small amount of time. Slow blow fuses have a coiled construction inside. They are designed to open only on a continued overload, such as a short circuit.

Circuit breakers

A circuit breaker is an automatically operated electrical switch designed to protect an electrical circuit from damage caused by an overload or short circuit. Figure 9-22 shows a circuit breaker panel. Its basic function is to detect a fault condition and immediately discontinue electrical flow. Unlike a fuse that operates once and then has to be replaced, a circuit breaker can be reset to resume normal operation. All resettable circuit breakers should open the circuit in which they are installed regardless of the position of the operating control when an overload or circuit fault exists. Such circuit breakers are referred to as trip-free. Automatic

Figure 9-22   Circuit breaker panel.

reset circuit breakers automatically reset themselves. They should not be used as circuit protection devices in aircraft. When a circuit breaker trips, the electrical circuit should be checked and the fault removed before the circuit breaker is reset. Sometimes circuit breakers trip for no apparent reason and the circuit breaker can be reset one time. If the circuit breaker trips again, there exists a circuit fault and the technician must troubleshoot the circuit before resetting the circuit breaker.

10

Aircraft Drawings

A *drawing* is a method to convey ideas concerning the construction or assembly of objects. This is done with the help of lines, notes, abbreviations, and symbols. It is very important that the aviation mechanic who is to make or assemble the object understand the meaning of the different lines, notes, abbreviations, and symbols that are used in a drawing.

Although blueprints as such are no longer used, the term *blueprint* or *print* is generally used in place of *drawing*.

Orthographic Projection

In order to show the exact size and shape of all the parts of complex objects, a number of views are necessary. This is the system used in orthographic projection.

Orthographic projection shows six possible views of an object because all objects have six sides: front, top, bottom, rear, right side, and left side. See Fig. 10-1.

It is seldom necessary to show all six views to portray an object clearly; therefore, only those views necessary to illustrate the required characteristics of the object are drawn. One-view, two-view, and three-view drawings are the most common.

Advanced computer-aided design (CAD) and computer-aided manufacturing (CAM) programs are used to generate drawings and computer numeric (NC) data and have replaced the traditional blueprint or print used in the past. These programs are

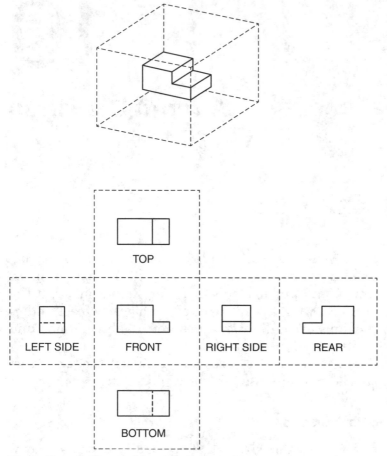

Figure 10-1   Orthographic projection.

used to design, analyze, test, and manufacture complete aircraft structures.

Working Drawings

Working drawings must give such information as size of the object and all of its parts, its shape and that of all of its parts, specifications as to the material to be used, how the material is to be finished, how the parts are to be assembled, and any other information essential to making and assembling the particular object.

Working drawings can be divided into three classes: detail, assembly, and installation.

Detail drawing

A detail drawing is a description of a single part, given in such a manner as to describe by lines, notes, and symbols the specifications as to size, shape, material, and methods of manufacture that are to be used in making the part. Detail drawings are usually rather simple and, when single parts are small, several detail drawings might be shown on the same sheet or print.

Assembly drawing

An assembly drawing is a description of an object consisting of two or more parts. It describes the object by giving, in a general way, the size and shape. Its primary purpose is to show the relationship of the various parts. An assembly drawing is usually more complex than a detail drawing, and is often accompanied by detail drawings of various parts.

Installation drawing

An installation drawing is one that includes all necessary information for a part or an assembly of parts in the final position in the aircraft. It shows the dimensions necessary for the location of specific parts with relation to the other parts and reference dimensions that are helpful in later work in the shop.

A pictorial drawing is similar to a photograph. It shows an object as it appears to the eye, but it is not satisfactory for showing complex forms and shapes. Pictorial drawings are useful in showing the general appearance of an object and are used extensively with orthographic projection drawings. Pictorial drawings are used in maintenance and overhaul manuals.

Title Block

All working drawings include a title block with the following information as shown in Fig. 10-2.

- The name of the company that produces the part.
- Number of the drawing. If it is a detail drawing, the drawing number is also the part number.

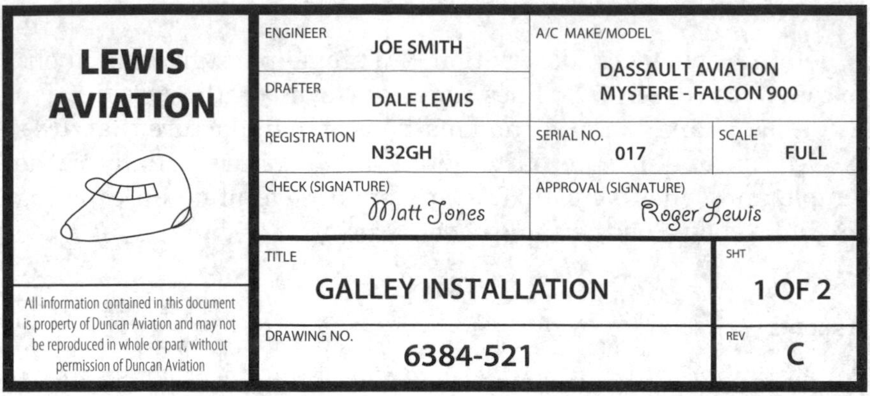

LEWIS AVIATION	ENGINEER JOE SMITH	A/C MAKE/MODEL	
	DRAFTER DALE LEWIS	DASSAULT AVIATION MYSTERE - FALCON 900	
	REGISTRATION N32GH	SERIAL NO. 017	SCALE FULL
	CHECK (SIGNATURE) *Matt Jones*	APPROVAL (SIGNATURE) *Roger Lewis*	
All information contained in this document is property of Duncan Aviation and may not be reproduced in whole or part, without permission of Duncan Aviation	TITLE GALLEY INSTALLATION		SHT 1 OF 2
	DRAWING NO. 6384-521		REV C

Figure 10-2 Title block.

- The scale to which it is drawn. Although a part is normally accurately drawn, the drawn part should not be scaled to obtain a dimension.
- The date of the finished drawing.
- The names and signatures of the draftsman, checker, and persons approving the drawing.
- If the drawing applies to an aircraft, the manufacturer's model number will be included.

Bill of Material

A list of the materials and parts necessary for the fabrication or assembly of a component or system is often included on the drawing. The list is usually in ruled columns in which are listed the part number, name of the part, material from which the part is to be constructed, the quantity required, and the source of the part or material. A typical bill of material is shown in Fig. 10-3. On drawings that do not have a bill of material, the data may be indicated directly on the drawing.

Other Data

Depending on the complexity of the items on the drawing, a revision block might be included to indicate any changes to the

BILL OF MATERIAL

ITEM	PART NO.	REQUIRED	SOURCE
CONNECTOR	UG-21 D/U	2	STOCK

Figure 10-3 Bill of material.

original. Notes are sometimes added for various clarifying reasons. Finish marks are used to indicate the surfaces that must be machine finished. Most dimensions will include tolerances or the total allowable variation of a size.

Sectional Views

A section or sectional view is obtained by cutting away part of an object to show the shape and construction at the cutting plane. The part or parts cut away are shown by the use of section (cross-hatching) lines as shown in Fig. 10-4.

Sectional views are used when the interior construction or hidden features of an object cannot be shown clearly by exterior views.

The Lines on a Drawing

Every drawing is composed of lines. Lines mark the boundaries, edges, and intersection of surfaces. Lines are used to show dimensions and hidden surfaces, and to indicate centers. Obviously, if the same kind of line is used to show all of these things, a drawing becomes a meaningless collection of lines. For this reason, various kinds of standardized lines are used on aircraft drawings, as shown in Fig. 10-5.

Most drawings use three widths or intensities of lines: thin, medium, or thick. These lines might vary somewhat on different

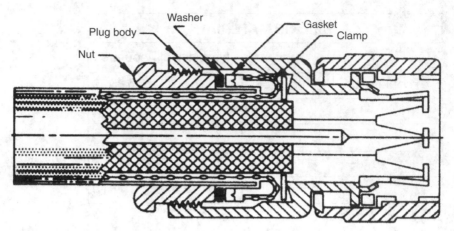

Figure 10-4   Sectional view of a cable connector.

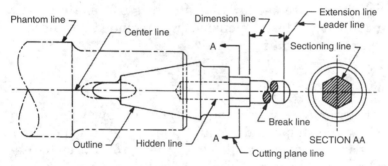

Figure 10-5   Example of correct use of lines.

drawings, but there will always be a noticeable difference between a thin and a thick line. The width of the medium line will be somewhere between the two. Figure 10-5 shows the correct use of lines by example.

Rivet Symbols Used on Drawings (Blueprints)

Rivet locations are shown on drawings by symbols. These symbols provide the necessary information by the use of code numbers or code letters or a combination of both. The meaning of the code

Figure 10-6 Basic rivet symbol quadrant configuration.

numbers and code letters is explained in the general notes section of the drawing on which they appear.

The rivet code system has been standardized by the National Aerospace Standards Committee (NAS Standard) and has been adopted by most major companies in the aircraft industry. This system has been assigned the number *NAS523* in the NAS Standard book.

The NAS523 basic rivet symbol consists of two lines crossing at 90°, which form four quadrants. Code letters and code numbers are placed in these quadrants to give the desired information about the rivet. Each quadrant has been assigned a name: *northwest (NW)*, *northeast (NE)*, *southwest (SW)*, and *southeast (SE)* (Fig. 10-6).

The rivet type, head type, size, material, and location are shown on the field of the drawing by means of the rivet code, with one exception. Rivets to be instated flush on both sides are not coded, but are called out and detailed on the drawing. An explanation of the rivet codes for each type of rivet used is shown on the field of the drawing. Figure 10-7 shows examples

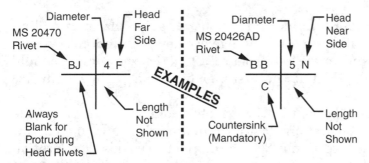

Figure 10-7 Examples of rivet coding on a drawing.

CODE	BASIC PART NO.	MATERIAL	DESCRIPTION OF RIVET
BA	MS 20426A	1100F	Solid, 100° Flush
BB	MS 20426AD	2117-T3	Solid, 100° Flush
CY	MS 20426DD	2024-T31	Solid, 100° Flush
BH	MS 20470A	1100F	Solid, Universal Head
BJ	MS 20470AD	2117-T3	Solid, Universal Head
CX	MS 20470DD	2024-T31	Solid, Universal Head
AAR	NAS 1738E	5056	Blind, Protruding Head
AAP	NAS 1738M	MONEL	Blind, Protruding Head
AAV	NAS 1739E	5056	Blind, 100° Flush
AAW	NAS 1739M	MONEL	Blind, 100° Flush

Figure 10-8 Typical examples of rivet coding. This list will vary according to requirements of each manufacturer.

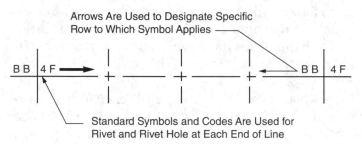

Figure 10-9 Method of illustrating rivet codes and the location where a number of identical rivets are in a row.

of rivet coding on the drawing and Fig. 10-8 is a sample of rivet coding.

Hole and countersink dimensions for solid-shank and blind rivets are omitted on all drawings because it is understood that the countersink angle is 100 degrees, and the countersink should be of such depth that the fastener fits flush with the surface after driving.

Where a number of identical rivets are in a row, the rivet code is shown for the first and last rivet in the row only, and an arrow will show the direction in which the rivet row runs. The location of the rivets between the rivet codes are marked only with crossing centerlines, as shown in Fig. 10-9.

11

Nondestructive Testing (NDT) or Nondestructive Inspection (NDI)

Unlike the previous chapters, which provided "hands on," detailed procedures for accomplishing a given task (such as drilling, riveting, etc.), this presentation of NDT is more general. Detailed procedures for using all of the NDT methods in use today are beyond the scope of this book. Therefore, a broad overview of each of the NDT methods is presented to familiarize the technician with the many variations of this important subject.

Visual Inspection

Visual inspection is the oldest of the nondestructive methods of testing. It is a quick and economical method to detect various types of cracks before they progress to failure. Its reliability depends upon the ability and experience of the inspector. He or she must know how to search for structural failures and how to recognize areas where such failures are likely to occur. Defects that would otherwise escape the naked eye can often be detected with the aid of optical devices.

The equipment necessary for conducting a visual inspection usually consists of a strong flashlight, a mirror with a ball joint, and a 2.5× – 4× magnifying glass. A 10× magnifying glass is recommended for positive identification of suspected cracks. Visual inspection of some areas can be made only with the use of a borescope.

Many borescopes provide images that can be displayed on a computer or video monitor for better interpretation of what is being viewed and to record images for future reference.

NDT Beyond Visual Inspection

Advanced NDT methods have rapidly evolved in the last decade and the introduction of composite structures has introduced several new NDT methods that are unique to composite structures. The five major methods for metallic aircraft are discussed first and followed by three NDT methods that have been developed for composite structures. The five most common methods used for metallic aircraft are: liquid penetrant, magnetic particle, eddy current, ultrasonic, and radiography. Additional NDT methods used for composite structures are: coin or tap test, thermography, and shearography. The field of NDT is constantly evolving and new methods are introduced frequently.

Liquid penetrant inspection

Penetrant inspection is a nondestructive test for defects open to the surface in parts made of any nonporous material. It is used with equal success on such metals as aluminum, magnesium, brass, copper, cast iron, stainless steel, and titanium. It may also be used on ceramics, plastics, molded rubber, and glass. Penetrant inspection

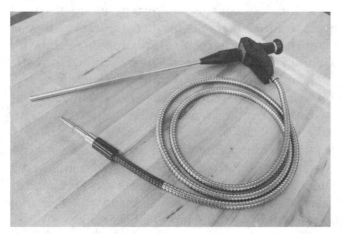

Figure 11-1 Borescope.

will detect such defects as surface cracks or porosity. These defects may be caused by fatigue cracks, shrinkage cracks, shrinkage porosity, cold shuts, grinding and heat-treat cracks, seams, forging laps, and bursts. Penetrant inspection will also indicate a lack of bond between joined metals.

The main disadvantage of penetrant inspection is that the defect must be open to the surface in order to let the penetrant get into the defect. For this reason, if the part in question is made of material which is magnetic, the use of magnetic particle inspection is generally recommended.

Penetrant inspection uses a penetrating liquid that enters a surface opening and remains there, making it clearly visible to the inspector. It calls for visual examination of the part after it has been processed, increasing the visibility of the defect so that it can be detected. Visibility of the penetrating material is increased by the addition of one of two types of dye, visible or fluorescent. There are two types of liquid penetrant kits available: a visible penetrant

Figure 11-2 Dye penetrant operation.

kit, which shows the defect directly; and a fluorescent kit, which contains a black light assembly to make the defect visible.

Eddy-current inspection

Electromagnetic analysis is a term which describes the broad spectrum of electronic test methods involving the intersection of magnetic fields and circulatory currents. The most widely used technique is the eddy current. Eddy currents are composed of free electrons under the influence of an induced electromagnetic field which are made to "drift" through metal. Figure 11-3 shows eddy-current inspection equipment.

Basic principles.   When an alternating current is passed through a coil, it develops a magnetic field around the coil, which in turn induces a voltage of opposite polarity in the coil and opposes the flow of original current. If this coil is placed in such a way that the magnetic field passes through an electrically conducting specimen, eddy currents will be induced into the specimen. The eddy currents create their own field which varies the original field's opposition to the flow of original current. The specimen's susceptibility to eddy currents determines the current flow through the coil. The magnitude and phase of this counter field is dependent primarily upon the resistance and permeability of the specimen under consideration, and which enables us to make a qualitative determination of various physical properties of the test material. The interaction of the eddy

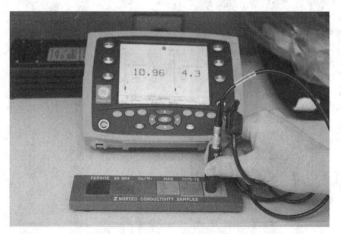

Figure 11-3   Eddy-current inspection equipment.

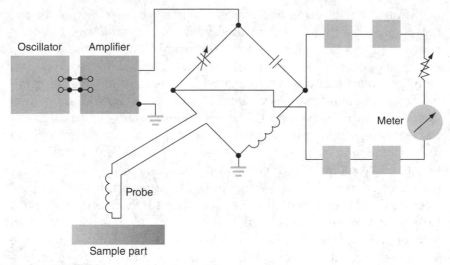

Figure 11-4 Eddy-current basic circuitry.

current field with the original field results is a power change that can be measured by utilizing electronic circuitry similar to a Wheatstone bridge as shown in Fig. 11-4. The specimen is either placed in or passed through the field of an electromagnetic induction coil, and its effect on the impedance of the coil or on the voltage output of one or more test coils is observed. The process, which involves electric fields made to explore a test piece for various conditions, involves the transmission of energy through the specimen much like the transmission of x-rays, heat, or ultrasound. Eddy-current inspection can frequently be performed without removing the surface coatings such as primer, paint, and anodized films. It can be effective in detecting surface and subsurface corrosion, pots, and heat-treat condition.

Ultrasonic inspection

Ultrasonic detection equipment makes it possible to locate defects in all types of materials. Ultrasonic methods are frequently used to inspect advanced composite structures. Minute cracks, checks, and voids too small to be seen by x-ray can be located by ultrasonic inspection. An ultrasonic test instrument requires access to only one surface of the material to be inspected and can be used with either straight line or angle-beam testing techniques.

Ultrasonic inspection has proven to be a very useful tool for the detection of internal delaminations, voids, or inconsistencies in composite components not otherwise discernable using visual or tap methodology. There are many ultrasonic techniques; however, each technique uses sound wave energy with a frequency above the audible range. A high-frequency (usually several MHz) sound wave is introduced into the part and may be directed to travel normal to the part surface, or along the surface of the part, or at some predefined angle to the part surface. You may need to try different directions to locate the flow. The introduced sound is then monitored as it travels its assigned route through the part for any significant change. Ultrasonic sound waves have properties similar to light waves. When an ultrasonic wave strikes an interrupting object, the wave or energy is either absorbed or reflected back to the surface. The disrupted or diminished sonic energy is then picked up by a receiving transducer and converted into a display on an oscilloscope or a chart recorder. The display allows the operator to evaluate the discrepant indications comparatively with those areas known to be good. To facilitate the comparison, reference standards are established and utilized to calibrate the ultrasonic equipment.

The two main methods of ultrasonic inspection are pulse-echo and through-transmission. Other methods such as phase array and bond tester are frequently used for composite structures. Figure 11-5 shows the basic through-transmission and pulse-echo methods.

Through-transmission ultrasonic inspection.   Through-transmission ultrasonic inspection uses two transducers, one on each side of the area to be inspected. The ultrasonic signal is transmitted from one transducer to the other transducer. The loss of signal strength is then measured by the instrument. The instrument shows the loss as a percent of the original signal strength or the loss in decibels. The signal loss is compared to a reference standard. Areas with a greater loss than the reference standard indicate a defective area.

Pulse-echo ultrasonic inspection.   Single-side ultrasonic inspection may be accomplished using pulse-echo techniques. In this method, a single search unit is working as a transmitting and a receiving transducer that is excited by high-voltage pulses. Each electrical pulse activates the transducer element. This element converts the electrical energy into mechanical energy

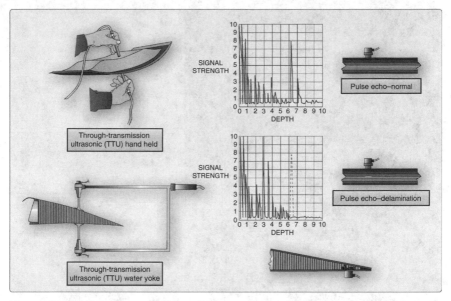

Figure 11-5 Pulse-echo and through-transmission inspection methods.

in the form of an ultrasonic sound wave. The sonic energy travels through a Teflon® or methacrylate contact tip into the test part. A waveform is generated in the test part and is picked up by the transducer element. Any change in amplitude of the received signal, or time required for the echo to return to the transducer, indicates the presence of a defect. Pulse-echo inspections are used to find cracks in metallic aircraft and delaminations, cracks, porosity, water, and disbonds in composite aircraft structures. Figure 11-6 shows pulse-echo test equipment.

Phased array inspection

Phased array inspection is one of the latest ultrasonic instruments to detect flaws in composite structures. It operates under the same principle of operation as pulse echo, but it uses 64 sensors at the same time, which speeds up the process. Figure 11-7 shows a phase array tester.

Magnetic particle inspection

Magnetic particle inspection is a method of detecting invisible cracks and other defects in ferromagnetic materials such as iron

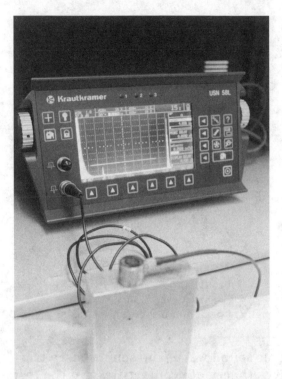

Figure 11-6   Pulse-echo test equipment.

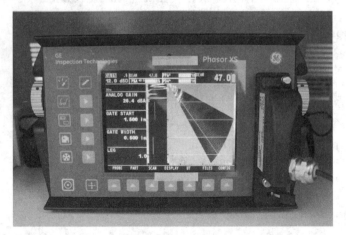

Figure 11-7   Phase array tester.

and steel. It is not applicable to nonmagnetic materials. The inspection process consists of magnetizing the part and then applying ferromagnetic particles to the surface area to be inspected. The ferromagnetic particles (indicating medium) may be held in suspension in a liquid that is flushed over the part; the part may be immersed in the suspension liquid; or the particles, in dry powder form, may be dusted over the surface of the part. The wet process is more commonly used in the inspection of aircraft parts. If a discontinuity is present, the magnetic lines of force will be disturbed and opposite poles will exist on either side of the discontinuity. The magnetized particles thus form a pattern in the magnetic field between the opposite poles. This pattern, known as an "indication," assumes the approximate shape of the surface projection of the discontinuity. A discontinuity may be defined as an interruption in the normal physical structure or configuration of a part, such as a crack, forging lap, seam, inclusion, porosity, and the like. The permanent magnetism remaining after inspection must be removed by a demagnetization operation if the part is to be returned to service. Figure 11-8 shows a fixed general-purpose magnetizing unit.

Radiography

X and gamma radiations, because of their unique ability to penetrate material and disclose discontinuities, have been applied to the radiographic (x-ray) inspection of metal fabrications and nonmetallic products. The penetrating radiation is projected through the part to be inspected and produces an invisible or latent image in the film. Figure 11-9 shows the basic principle of radiography. When processed, the film becomes a radiograph or shadow picture of the object. The traditional film is now often replaced by a digital image. This inspection medium and portable unit provides a fast and reliable means for checking the integrity of airframe structures and engines. Operators should always be protected by sufficient lead shields, as the possibility of exposure exists either from the x-ray tube or from scattered radiation. Maintaining a minimum safe distance from the x-ray source is always essential.

The following three NDT methods are used extensively on composite structures. Ultrasonic methods are also often used. Figure 11-10 shows what type of NDT method could be used to inspect composite structures.

(a)

(b)

Figure 11-8 (a) Fixed general-purpose magnetizing unit. (b) Magnetic particle inspection in progress.

Tap or coin test

Sometimes referred to as audio, sonic, or coin tap, this technique makes use of frequencies in the audible range (10 Hz to 20 Hz). A surprisingly accurate method in the hands of experienced personnel, tap testing is perhaps the most common technique used for the detection of delamination and/or disbonds in composite materials. The method is accomplished by tapping the inspection area with a solid round disk or lightweight hammer-like device

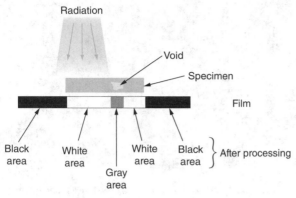

Figure 11-9 Radiography.

and listening to the response of the structure to the hammer, see Fig. 11-11. A clear, sharp, ringing sound is indicative of a well-bonded solid structure, while a dull or thud-like sound indicates a discrepant area.

Thermography

Thermography is a popular inspection method to inspect composite honeycomb structures. Thermal inspection comprises all methods in which heat-sensing devices are used to measure temperature variations for parts under inspection. The basic principle of thermal inspection consists of measuring or mapping of surface temperatures when heat flows from, to, or through a test object. All thermographic techniques rely on differentials in thermal conductivity between normal, defect-free areas, and those

Method of Inspection	Type of Defect							
	Disbond	Delamination	Dent	Crack	Hole	Water Ingestion	Overheat and Burns	Lightning Strike
Visual	X (1)	X (1)	X	X	X		X	X
X-Ray	X (1)	X (1)		X (1)		X		
Ultrasonic TTU	X	X						
Ultrasonic pulse echo		X				X		
Ultrasonic bondtester	X	X						
Tap test	X (2)	X (2)						
Infrared thermography	X (3)	X (3)				X		
Dye penetrant				X (4)				
Eddy current				X (4)				
Shearography	X (3)	X (3)						

Notes: (1) For defects that open to the surface
        (2) For thin structure (3 plies or less)
        (3) The procedures for this type of inspection are being developed
        (4) This procedure is not recommended

Figure 11-10 Comparison of NDT methods for inspection of composite structures.

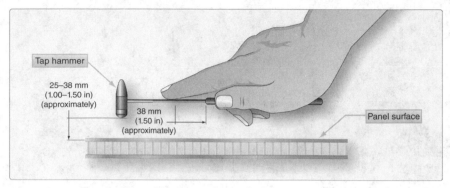

Figure 11-11   Tap testing of honeycomb sandwich structure with tap hammer.

having a defect. Normally, a heat source is used to elevate the temperature of the part being examined while observing the surface heating effects. Because defect-free areas conduct heat more efficiently than areas with defects, the amount of heat that is either absorbed or reflected indicates the quality of the bond. The type of defects that affect the thermal properties include debonds, cracks, impact damage, panel thinning, and water ingress into composite materials and honeycomb core. Thermal methods are most effective for thin laminates or for defects near the surface.

Shearography

Shearography is an optical NDI technique that detects defects by measuring the variations in reflected light (speckle pattern) from the surface of the object. Using a laser light source, an original image of the illuminated surface is recorded via a video image. The part is subsequently stressed by heating, changes in pressure or acoustic vibrations during which a second video image is made. Changes in the surface contour caused by disbonding or delaminating become visible on the video display. Shearography is being used in production environments for rapid inspection of bonded composite structure assemblies including carbon/epoxy skin and Nomex core sandwiches. This is accomplished by inducing stresses by partial vacuum. Partial vacuum stressing causes air content defects to expand, leading to slight surface deformations that are detected before and during stressing comparisons. Display of the computer-processed video image comparisons reveals defects as bright and dark concentric circles of constructive and destructive reflected light wave interference.

12

Corrosion Detection
and Control

Metal corrosion is the deterioration of the metal by chemical or electrochemical attack and can occur internally, as well as on the surface. This deterioration may change the smooth surface, weaken the interior, or damage or loosen adjacent parts.

Water or water vapor containing salt combines with oxygen in the atmosphere to produce the main source of corrosion in aircraft. Aircraft operating in a marine environment or in areas where the atmosphere contains corrosive industrial fumes are particularly susceptible to corrosive attacks.

Corrosion can cause eventual structural failure if left unchecked. The appearance of the corrosion varies with the metal. On aluminum alloys and magnesium, it appears as surface pitting and etching, often combined with a grey or white powdery deposit. On steel, it forms a reddish rust. When the grey, white, or reddish deposits are removed, each of the surfaces might appear etched and pitted, depending on the length of exposure and the severity of attack. If these surface pits are not too deep, they might not significantly alter the strength of the metal; however, the pits might become sites for crack development. Some types of corrosion can travel beneath surface coatings and can spread until the part fails.

Types of Corrosion

Two general classifications of corrosion, direct chemical attack and electrochemical attack, cover most of the specific forms. In both types of corrosion, the metal is converted into a metallic compound, such as an oxide, hydroxide, or sulfate. The corrosion process always involves two simultaneous changes: The metal that is attacked or oxidized suffers what might be called *anodic change*, and the corrosive agent is reduced and might be considered as undergoing a *cathodic change*.

Direct Chemical Attack

Direct chemical attack, or pure chemical corrosion, is an attack that results from a direct exposure of a bare surface to caustic liquid or gaseous agents. Unlike electrochemical attack, where the anodic and cathodic changes might be occurring a measurable distance apart, the changes in direct chemical attack are occurring simultaneously at the same point. The most common agents causing direct chemical attack on aircraft are:

- Spilled battery acid or fumes from batteries.
- Residual flux deposits resulting from inadequately cleaned, welded, brazed, or soldered joints.
- Entrapped caustic cleaning solutions.

Spilled battery acid is becoming less of a problem with the advent of aircraft using nickel-cadmium batteries, which are usually closed units.

Electrochemical Attack

The electrochemical attack is responsible for most forms of corrosion on aircraft structure and component parts.

An electrochemical attack can be likened chemically to the electrolytic reaction that occurs in electroplating, anodizing, or in a dry-cell battery. The reaction in this corrosive attack requires a medium, usually water, which is capable of conducting a tiny current of electricity. When a metal comes in contact with a corrosive agent and is also connected by a liquid or gaseous path through which electrons flow, corrosion begins as the metal decays by oxidation. During the attack, the quantity of corrosive agent is reduced and, if not renewed or removed, might completely react with the metal (become neutralized). Different areas of the same metal surface have varying levels of electrical potential and, if connected by a conductor, such as salt water, will set up a series of corrosion cells so that corrosion will commence as shown in Fig. 12-1.

All metals and alloys are electrically active and have a specific electrical potential in a given chemical environment as shown in Fig. 12-2. The constituents in an alloy also have specific electrical potentials that are generally different from each other. Exposure of the alloy surface to a conductive, corrosive medium causes the more active metal to become anodic and the less-active metal to become cathodic, thereby establishing conditions for corrosion. These are called *local cells*. The greater the difference in electrical

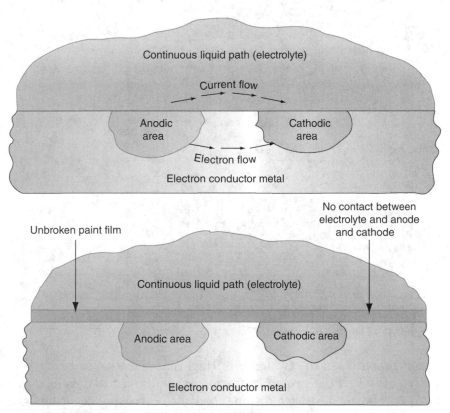

Figure 12-1 Electrochemical corrosion attack.

+ Corroded End (anodic, or least noble)
Magnesium Magnesium alloy Zinc
Aluminum (1100) Cadmium Aluminum 2024-T4 Steel or Iron Cast Iron Chromium-Iron (active) Ni-Resist Cast Iron
Type 304 Stainless steel (active) Type 316 Stainless steel (active)
Lead-Tin solder Lead Tin
Nickel (active) Inconel nickel-chromium alloy (active) Hastelloy Alloy C (active)
Brass Copper Bronze Copper-nickel alloy Monel nickel-copper alloy
Silver Solder Nickel (passive) Inconel nickel-chromium alloy (passive)
Chromium-Iron (passive) Type 304 Stainless steel (passive) Type 316 Stainless steel (passive) Hastelloy Alloy C (passive)
Silver Titanium Graphite Gold Platinum
– Protected End (cathodic, or most noble)

Figure 12-2 The galvanic series of metals and alloys.

potential between the two metals, the greater the severity of a corrosive attack, if the proper conditions are allowed to develop.

As can be seen, the conditions for these corrosive reactions are a conductive fluid and metals having a difference in potential.

If, by regular cleaning and surface refinishing, the medium is removed and the minute electrical circuit is eliminated, corrosion cannot occur; this is the basis for effective corrosion control.

Forms of Corrosion

Surface corrosion

Surface corrosion appears as a general roughening, etching, or pitting of the surface of a metal, frequently accompanied by a powdery deposit of corrosion products.

Filiform corrosion

Filiform corrosion gives the appearance of a series of small worms under the paint surface. It is often seen on surfaces that have been improperly chemically treated prior to painting. Figure 12-3 shows filiform corrosion.

Pitting corrosion

Extensive pitting damage, as shown in Fig. 12-4, may result from contact between dissimilar metal parts in the presence of a conductor. While surface corrosion may or may not be taking place, a galvanic action, not unlike electroplating, occurs at the points or areas of contact where the insulation between the surfaces has broken down or been omitted.

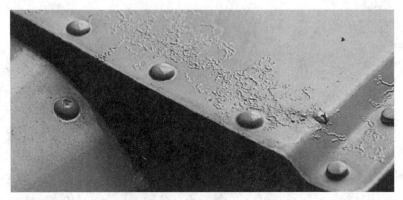

Figure 12-3   Filiform corrosion.

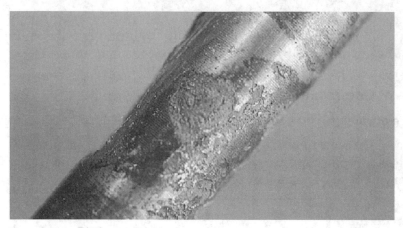

Figure 12-4 Pitting corrosion.

Intergranular corrosion

This type of corrosion is an attack along the grain boundaries of an alloy and commonly results from a lack of uniformity in the alloy structure. Aluminum alloys and some stainless steels are particularly susceptible to this form of electrochemical attack; see Fig. 12-5. The lack of uniformity is caused by changes that occur in the alloy during heating and cooling during the material's manufacturing process.

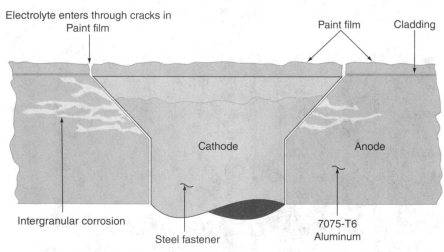

Figure 12-5 Intergranular corrosion of 7075-T6 alloy.

Figure 12-6 Exfoliation corrosion.

Exfoliation corrosion

Very severe intergranular corrosions as shown in Fig. 12-6, may sometimes cause the surface of a metal to "exfoliate." This is a lifting or flaking of the metal at the surface due to delamination of the grain boundaries caused by the pressure of corrosion residual product buildup.

Stress corrosion

Stress corrosion occurs as the result of the combined effect of sustained tensile stresses and a corrosive environment. Stress corrosion cracking is found in most metal systems; however, it is particularly characteristic of aluminum, copper, certain stainless steels, and high-strength alloy steels (over 240,000 psi).

Fretting corrosion

Fretting corrosion occurs when two mating surfaces, normally at rest with respect to one another, are subject to slight relative motion (Fig. 12-7). It is characterized by pitting of the surfaces and the generation of considerable quantities of finely divided debris.

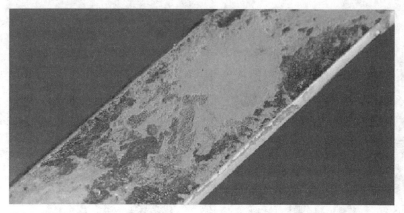

Figure 12-7   Fretting corrosion.

Effects of Corrosion

Most metals are subject to corrosion, but corrosion can be minimized by use of corrosion-resistant metals and finishes. The principal material used in air-frame structures is high-strength aluminum alloy sheet coated (clad) with a pure aluminum coating (alclad), which is highly resistant to corrosive attack. However, with an accumulation of airborne salts and/or industrial pollutants with an electrolyte (moisture), pitting of the alclad will occur. Once the alclad surface is broken, rapid deterioration of the high-strength aluminum alloy below occurs. Other metals commonly used in air-frame structure, such as nonclad high-strength aluminum alloys, steel, and magnesium alloys, require special preventive measures to guard against corrosion. The characteristics of corrosion in commonly used aircraft metals is summarized in Fig. 12-8.

The degree of severity, the cause, and the type of corrosion depend on many factors, including the size or thickness of the part, the material, heat treatment of the material, protective finishes, environmental conditions, preventative measures, and design. Thick structural sections are generally more susceptible to corrosive attack because of variations in their composition, particularly if the sections are heat treated during fabrication.

Corrosion Control

Nearly any durable coating that creates a moisture barrier between a metal substrate and the environment will help control

Alloy	Type of Attack to Which Alloy is Susceptible	Appearance of Corrosion Product
Magnesium	Highly susceptible to pitting	White, powdery, snowlike mounds and white spots on surface
Low Alloy Steel (4000–8000 series)	Surface oxidation and pitting, surface, and intergranular	Reddish-brown oxide (rust)
Aluminum	Surface pitting, intergranular, exfoliation stresscorrosion and fatigue cracking, and fretting	White-to-grey powder
Titanium	Highly corrosion resistant; extended or repeated contact with chlorinated solvents may result in degradation of the metal's structural properties at high temperature	No visible corrosion products at low temperature. Colored surface oxides develop above 700°F (370°C)
Cadmium	Uniform surface corrosion; used as sacrificial plating to protect steel	From white powdery deposit to brown or black mottling of the surface
Stainless steels (300–400 series)	Crevice corrosion; some pitting in marine environments; corrosion cracking; intergranular corrosion (300 series); surface corrosion (400 series)	Rough surface; sometimes a uniform red, brown, stain

Figure 12-8   Results of corrosion attack on metals.

or prevent corrosion. Paints, waxes, lubricants, water-displacing compounds, penetrating oils, or other hard or soft coatings can provide an effective moisture barrier.

Exposure to marine atmosphere, moisture, acid rain, tropical temperature conditions, industrial chemicals, and soils and dust in the atmosphere contribute to corrosion. Limit, whenever possible, the requirement for operation of aircraft in adverse environments.

Corrosion preventive compounds, such as LPS Procyon, Dinol, Zip-Chem (or equivalent products), and later advanced developments of such compounds, can be used to effectively reduce the occurrence of corrosion. Results of corrosion inspections should be reviewed to help establish the effectiveness of corrosion-preventive compounds and determine the reapplication interval of them (see Fig. 12-9).

Figure 12-9 Corrosion inhibitor compounds are available in aerosol cans for touch-up or 1-gallon, 5-gallon, and 55-gallon containers for spraying with a wand. These compounds produce a transparent film, resistant to salt spray, moisture, and most typical corrosive elements. Courtesy LPS Laboratories, Inc.

Inspection Requirements

Except for special requirements in trouble areas, inspection for corrosion should be a part of routine maintenance inspections. Trouble areas, however, are a different matter, and experience shows that certain combinations of conditions result in corrosion in spite of routine inspection requirements. These trouble areas might be peculiar to particular aircraft models, but similar conditions are usually found on most aircraft. Most manufacturers' handbooks of inspection requirements are complete enough to cover all parts of the aircraft or engine, and no part or area of the aircraft should go unchecked. Use these handbooks as a general guide when an area is to be inspected for corrosion.

Corrosion Prevention

Corrosion preventive maintenance includes the following specific functions:

1. Adequate cleaning
2. Thorough periodic lubrication
3. Detailed inspection for corrosion and failure of protective systems
4. Prompt treatment of corrosion and touchup of damaged paint areas
5. Keeping drain holes free of obstructions
6. Daily draining of fuel cell sumps

7. Daily wipe down of exposed critical areas

8. Sealing of aircraft against water during foul weather and proper ventilation on warm, sunny days

9. Maximum use of protective covers on parked aircraft

Corrosion-Prone Areas

Exhaust trail areas

Battery compartments and battery

Vent openings

Bilge areas

Wheel well and landing gear

Water entrapment areas

Engine frontal areas and cooling air vents

Wing flap and spoiler recesses

External skin areas

Corrosion-Removal Techniques

When active corrosion is apparent, a positive inspection and rework program is necessary to prevent any further deterioration of the structure. The following methods of assessing corrosion damage and procedures for reworking corroded areas could be used during the cleanup programs. In general, any rework could involve the cleaning and stripping of all finish from the corroded area, the removal of corrosion products, and restoration of surface-protective film.

The repair of corrosion damage includes removing all corrosion and corrosion products. When the corrosion damage exceeds the damage limits set by the aircraft manufacturer in the structural repair manual, the affected part must be replaced or an FAA-approved engineering authorization for continued service for that part must be obtained.

If the corrosion damage on large structural parts is in excess of that allowed in the structural repair manual and where replacement is not practical, contact the aircraft manufacturer for rework limits and procedures.

SUPERFICIAL CORROSION ON CLAD
OR NON-CLAD ALUMINUM ALLOY
SHEET, ALUMINUM ALLOY TUBING
AND ALUMINUM ALLOY
FORGING AND CASTINGS.

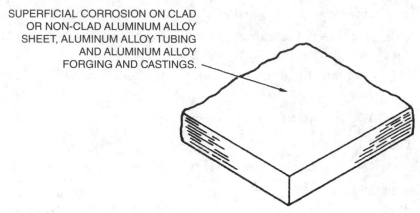

Figure 12-10 Repair of superficial surface corrosion on clad or non-clad aluminum alloy sheet.

Several standard methods are available for corrosion removal. The methods normally used to remove corrosion are mechanical and chemical. Mechanical methods include hand sanding using abrasive mat, abrasive paper, or metal wool; and powered mechanical sanding, grinding, and buffing, using abrasive mat, grinding wheels, sanding discs, and abrasive rubber mats. However, the method used depends upon the metal and the degree of corrosion.

Detailed procedures for removing corrosion and evaluating the damage are beyond the scope of this book.

Surface Damage by Corrosion

To repair of superficial corrosion on clad or non-clad aluminum alloy sheet, use the following procedure (see Fig. 12-10).

1. Remove corrosion from aluminum alloy sheet by the following methods:

   Non-clad   #400 sandpaper and water.

   Clad   Abrasive metal polish.

2. Apply 5% solution by weight of chromic acid after cleanup. Rinse with tap water to remove any chromic acid stains.

13

Composites

Introduction

New-generation aircraft are often designed with all composite fuselage and wing structures, and the repair of these advanced composite materials requires an in-depth knowledge of composite structures, materials, and tooling. The primary advantages of composite materials are their high strength, relatively low weight, fatigue resistance, and corrosion resistance.

Definition of Composite Materials

Composite materials consist of a combination of materials that are mixed together to achieve specific structural properties. The individual materials do not dissolve or merge completely in the composite, but they act together as one. Normally, the components can be physically identified as they interface with one another. The properties of the composite material are superior to the properties of the individual materials from which it is constructed.

Major Components of a Laminate

An advanced composite material is made of a fibrous material embedded in a resin matrix, generally laminated with fibers oriented in alternating directions to give the material strength and stiffness. Metals are isotropic materials, which means that they have uniform properties in all directions. Composite materials are anisotropic, which means that they have predominant properties in one

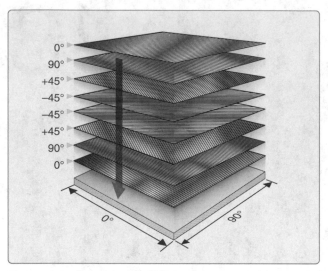

Figure 13-1   Quasi-isotropic layup.

direction (fiber direction). Many composite structures are quasi-isotropic, which means that the plies of the layup are stacked in a 0-, −45-, 45-, and 90-degree sequence or in a 0-, −60-, and 60-degree sequence. Figure 13-1 shows a quasi-isotropic layup.

Types of Fiber

Fiberglass

Fiberglass is often used for secondary structure on aircraft, such as fairings, radomes, and wing tips. Fiberglass is also used for helicopter rotor blades. Two types of fiberglass are used: E-glass, the most commonly used type of fiberglass; and S glass, which has higher mechanical properties. Fiberglass is identified by its white color.

Carbon

Carbon fibers are very stiff and strong, 3 to 10 times stiffer than glass fibers. Carbon fiber is used for structural aircraft applications, such as floor beams, stabilizers, flight controls, and primary fuselage and wing structure. Advantages include its high strength and corrosion resistance. Disadvantages include lower conductivity than aluminum; therefore, a lightning protection mesh or coating is necessary for aircraft parts that are prone to lightning strikes. Another disadvantage of carbon fiber is its high cost. Carbon fiber is gray or black in color.

Kevlar®

Kevlar® is DuPont's name for aramid fibers. Aramid fibers are lightweight, strong, and tough. Two types of aramid fiber are used in the aviation industry: Kevlar® 49 and Kevlar® 29. Kevlar® 49 has a high stiffness and Kevlar® 29 has a low stiffness. An advantage of aramid fibers is their high resistance to impact damage, so they are often used in areas prone to impact damage. The main disadvantage of aramid fibers is their general weakness in compression and its ability to absorb up to 8 percent of moisture. Kevlar® is difficult to drill and cut. The fibers fuzz easily and special scissors, drill bits, and countersinks are needed to cut the material.

Fiber Forms

All product forms generally begin with spooled unidirectional raw fibers packaged as continuous strands. An individual fiber is called a filament. The word strand is also used to identify an individual glass fiber. Bundles of filaments are identified as tows, yarns, or rovings. Fiberglass yarns are twisted, while Kevlar® yarns are not. Tows and rovings do not have any twist.

Roving

A roving is a single grouping of filament or fiber ends, such as 20-end or 60-end glass rovings. All filaments are in the same direction and they are not twisted.

Unidirectional (tape)

Unidirectional materials, also called tape, have all the fibers in one direction. The fibers are either stitched or preimpregnated with a resin so that they stay together. Unidirectional products

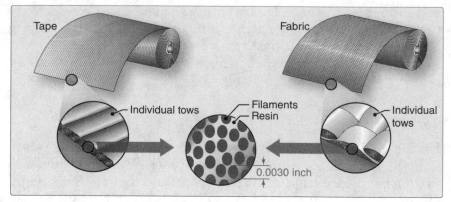

Figure 13-2 Tape and fabric products.

have high strength in the fiber direction and virtually no strength across the fibers; see Fig.13-2.

Bidirectional (fabric)

Bidirectional materials, also called fabrics, are made by a weaving process. Most fabric constructions offer more flexibility for layup of complex shapes than straight unidirectional tapes offer. The more common fabric styles are plain or satin weaves. The plain weave construction results from each fiber alternating over and then under each intersecting strand (tow, bundle, or yarn). With the common satin weaves, such as five harness or eight harness, the fiber bundles traverse both in warp and fill directions changing over/under position less frequently. Satin weaves are used when more fabric drape is required. Figure 13-3 shows various fabric weave styles.

Resin Systems

The resin system and its chemical composition and physical properties fundamentally affect the processing, fabrication, and ultimate properties of a composite material. Thermosetting resins are the most diverse and widely used of all man-made materials. They are easily poured or formed into any shape, are compatible with most other materials, and cure readily (by heat or catalyst) into an insoluble solid. Thermosetting resins are also excellent adhesives and bonding agents. Both thermoset and thermoplastic

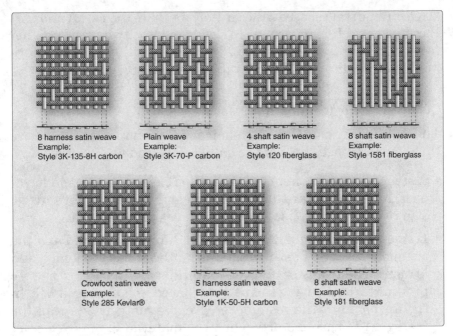

Figure 13-3 Various fabric weave styles.

resin systems are used in the aviation industry. The most commonly used thermoset resin systems are epoxy, polyester, vinyl ester, phenolic resin, bismaleimides (BMI), and polyimides. Thermoplastic materials can be softened repeatedly by an increase of temperature and hardened by a decrease in temperature. Processing speed is the primary advantage of thermoplastic materials. Chemical curing of the material does not take place during processing, and the material can be shaped by molding or extrusion when it is soft. Polyether ether ketone, better known as PEEK, is a high-temperature thermoplastic. This aromatic ketone material offers outstanding thermal and combustion characteristics and resistance to a wide range of solvents and proprietary fluids. PEEK can also be reinforced with glass and carbon.

Mixing two-part resin systems

Thermosetting resin systems are either one- or two-part systems. The one-part systems are already mixed and need to be stored in a freezer to retard curing. Two-part systems do not start curing until they are mixed. The mixing ratio is most often by weight and an electronic scale is used to accurately measure the part A and part B components. It is important to slowly mix the two parts together to avoid introducing air bubbles. Always consult the manufacturer's instructions for correct mixing ratios, pot life, and cure cycle.

Curing stages of thermosetting resins

Thermosetting resins use a chemical reaction to cure. There are three curing stages, which are called A, B, and C.

- A stage: The components of the resin (base material and hardener) have been mixed but the chemical reaction has not started. The resin is in the A stage during a wet layup procedure.

- B stage: The components of the resin have been mixed and the chemical reaction has started. The material has thickened and is tacky. The resins of prepreg materials are in the B stage. To prevent further curing the resin is placed in a freezer at 0°F. In the frozen state, the resin of the prepreg material stays in the B stage. The curing starts when the material is removed from the freezer and warmed again.

- C stage: The resin is fully cured. Some resins cure at room temperature and others need an elevated temperature cure cycle to fully cure.

Dry Fiber and Prepreg

Composite structures are made of either dry fiber or prepreg products. Dry fiber is impregnated with a resin system during the manufacturing process. Processes that use dry fiber are wet hand layup, filament winding, vacuum-assisted resin transfer molding (VARTM), and resin transfer molding (RTM). Another process that is often used is called prepreg. Prepreg material consists of a combination of a resin and fiber reinforcement. It is available in unidirectional form and fabric form as shown in Fig. 13-4. The resin is then no longer in a low-viscosity stage, but has been advanced to a B-stage level of cure for better handling characteristics. The following products are available in prepreg form: unidirectional tapes, woven fabrics, continuous strand rovings, and chopped mat. Prepreg materials must be stored in a freezer at a temperature below 0°F to retard the curing process. Prepreg materials are cured

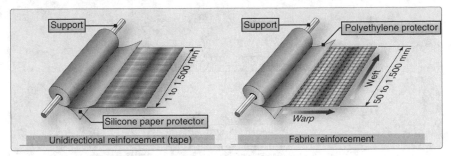

Figure 13-4   Prepreg tape and fabric products.

with an elevated temperature. Many prepreg materials used in aerospace are impregnated with an epoxy resin, and they are cured at either 250°F or 350°F. Prepreg products are used with hand layup, automated tape laying, automated fiber placement, and filament winding.

Adhesives

Three types of adhesives are used for the manufacture and repair of aircraft structures: film adhesives, paste adhesives, and foaming adhesives. Both one- and two-part systems are used.

Film adhesives

Structural adhesives for aerospace applications are generally supplied as thin films supported on a release paper and stored under refrigerated conditions (−18°C, or 0°F). These products have often a limited shelf life of only 6 months to 1 year. Rubber-toughened epoxy film adhesives are widely used in aircraft industry. Figure 13-5 shows typical applications for film adhesives.

Paste adhesives

Paste adhesives are used as an alternative to film adhesive. These are often used to secondary bond repair patches to damaged parts and also used in places where film adhesive is difficult to apply. Paste adhesives for structural bonding are made mostly from epoxy. One- and two-part systems are available. The advantages of paste adhesives are that they can be stored at room temperature and have a long shelf life.

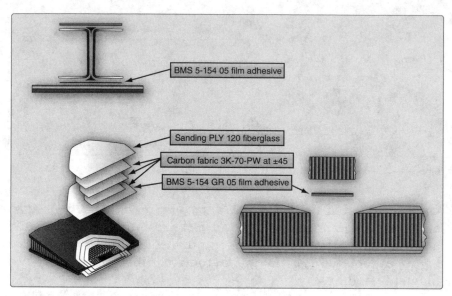

Figure 13-5   The application of film adhesives.

Foaming adhesives

Most foaming adhesives are 0.025- to 0.10-inch-thick sheets of B-staged epoxy. Foam adhesives cure at 250°F or 350°F. During the cure cycle, the foaming adhesives expand. Foaming adhesives need to be stored in the freezer just like prepregs and film adhesives, and they have only a limited storage life. Foaming adhesives are used to splice pieces of honeycomb together in a sandwich construction and to bond repair plugs to the existing core during a prepreg repair.

Honeycomb Sandwich Structures

A sandwich construction is a structural panel concept that consists in its simplest form of two relatively thin, parallel face sheets bonded to and separated by a relatively thick, lightweight core as shown in Fig. 13-6.

Increasing the core thickness greatly increases the stiffness of the honeycomb construction, while the weight increase is minimal. Due to the high stiffness of a honeycomb construction, it is

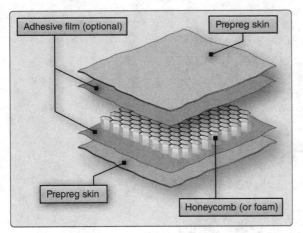

Figure 13-6 Honeycomb sandwich structure.

not necessary to use external stiffeners, such as stringers and frames; see Fig. 13-7. Most common face materials are carbon, fiberglass, and Kevlar®, and either aluminum or Nomex core materials are used. Typical aircraft structures that utilize honeycomb sandwich are radomes, fuselage to wing fairings, engine cowlings, flight controls, and spoilers.

Honeycomb core cells for aerospace applications are usually hexagonal for relatively flat panels. Overexpanded core is used for panels with simple curves, and bell-shaped core, or flexcore, is used for panels with complex shapes, as shown in Fig. 13-8.

	Solid Material	Core Thickness t	Core Thickness 3t
	t	2t	4t
Thickness	1.0	7.0	37.0
Flexural Strength	1.0	3.5	9.2
Weight	1.0	1.03	1.06

Figure 13-7 Core thickness greatly increases stiffness while hardly adding weight.

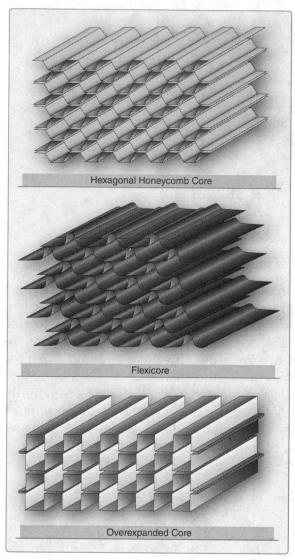

Figure 13-8 Types of honeycomb core.

Laminate Structures

Transport category aircraft use most often laminate structures instead of honeycomb sandwich for their primary fuselage and wing structure. Laminated structures have only one face sheet, which is externally reinforced with stiffeners, stringers, frames,

Figure 13-9 Laminate skin panel with reinforcing hat-section stringers.

and bulkheads to achieve a stiff structure. This type of structure is very similar to a semimonocoque metal aircraft construction. The laminated panel concept typically uses integrated stringers and stiffeners to reduce the part count; see Fig. 13-9. The advantages of a laminate structure compared to a honeycomb structure are increased impact resistance, less prone to moisture ingress, easier to repair, and the structure can be assembled using fasteners. The disadvantage is that a laminate structure is heavier than a honeycomb sandwich structure.

Damage and Defects

Manufacturing damage includes anomalies, such as porosity, microcracking, and delaminations resulting from processing discrepancies. It also includes such items as inadvertent edge cuts, surface gouges and scratches, damaged fastener holes, and impact damage. Examples of flaws occurring in manufacturing include a contaminated bondline surface or inclusions, such as prepreg backing paper or separation film that is inadvertently left between plies during layup. Inadvertent (nonprocess) damage can occur in detail parts or components during assembly or transport or during operation.

Delamination and debonds

Delaminations form on the interface between the layers in the laminate. Delaminations may form from matrix cracks that grow into the interlaminar layer or from low-energy impact. Debonds can also form from production nonadhesion along the bondline between two elements and initiate delamination in adjacent laminate layers.

Resin rich or starved

A part is resin rich if too much resin is used; for nonstructural applications this is not necessarily bad, but it adds weight. A part is called resin starved if too much resin is bled off during the curing process or if not enough resin is applied during the wet layup process. Resin-starved areas are indicated by fibers that show to the surface. The 60:40 fiber to resin ratio is considered optimal.

Fiber breakage

Fiber breakage can be critical because structures are typically designed to be fiber dominant (i.e., fibers carry most of the loads). Fortunately, fiber failure is typically limited to a zone near the point of impact and is constrained by the impact object size and energy.

Matrix imperfections

Matrix imperfections usually occur on the matrix-fiber interface or in the matrix parallel to the fibers. These imperfections can slightly reduce some of the material properties but are seldom critical to the structure, unless the matrix degradation is widespread.

Moisture ingress

Moisture ingress is a problem with composite structures, especially honeycomb sandwich structure. Composite materials absorb moisture and this moisture will affect the mechanical properties of the structure. Honeycomb panels could collect moisture inside the panel, which add weight and the moisture can also cause debonding to the structure when it freezes at high altitude or is heated to temperatures above 212°F during repair of the panel. The ingress of oil and hydraulic fluid could cause major damage to the core of the honeycomb core. If aluminum core is used corrosion of the core could be a major problem.

Vacuum Bagging Techniques

Repairs of composite aircraft components are often performed with a technique known as vacuum bagging. A plastic bag is sealed around the repair area. Air is then removed from the bag, which allows repair plies to be drawn together with no air trapped in between. Atmospheric pressure bears on the repair and a strong, secure bond is created. Several processing materials are used for vacuum bagging a part. These materials do not become part of the repair and are discarded after the repair process.

Release agents

Release agents, also called mold release agents, are used so that the part comes off the tool or caul plate easily after curing.

Bleeder ply

The bleeder ply creates a path for the air and volatiles to escape from the repair. Excess resin is collected in the bleeder. Bleeder material could be made of a layer of fiberglass, nonwoven polyester, or it could be a perforated Teflon®-coated material.

Peel ply

Peel plies are often used to create a clean surface for bonding purposes. A thin layer of fiberglass is cured with the repair part. Just before the part is bonded to another structure, the peel ply is removed. The peel ply is easy to remove and leaves a clean surface for bonding.

Layup tapes

Vacuum bag sealing tape, also called sticky tape, is used to seal the vacuum bag to the part or tool.

Perforated release film

Perforated parting film is used to allow air and volatiles out of the repair, and it prevents the bleeder ply from sticking to the part or repair. It is available with different size holes and hole spacing depending on the amount of bleeding required.

Solid release film

Solid release films are used so that the prepreg or wet layup plies do not stick to the working surface or caul plate. Solid release film is also used to prevent the resins from bleeding through and damaging the heat blanket or caul plate if they are used.

Breather material

The breather material is used to provide a path for air to get out of the vacuum bag. The breather must contact the bleeder. Typically, polyester is used in either 4- or 10-ounce weights. Four ounces is used for applications below 50 pounds per square inch (psi) and 10 ounces is used for 50 to 100 psi.

Vacuum bag

The vacuum bag material provides a tough layer between the repair and the atmosphere. The vacuum bag material is available in different temperature ratings, so make sure that the material used for the repair can handle the cure temperature. Most vacuum bag materials are one time use, but material made from flexible silicon rubber is reusable.

Curing and Curing Equipment

Some lower performance resin systems will cure at room temperature but most high-performance thermoset resin systems need an elevated temperature to cure. Autoclaves, curing ovens, and heat bonders are used to cure these resin systems.

Oven

Composite materials can be cured in ovens using various pressure application methods. Typically, vacuum bagging is used to remove volatiles and trapped air and utilizes atmospheric pressure for consolidation. Another method of pressure application for oven cures is the use of shrink wrapping or shrink tape. The oven uses heated air circulated at high speed to cure the material system. Typical oven cure temperatures are 250°F and 350°F.

Autoclave

An autoclave system allows a complex chemical reaction to occur inside a pressure vessel according to a specified time, temperature, and pressure profile in order to process a variety of materials. Modern autoclaves are computer controlled and the operator can write and monitor all types of cure cycle programs. Most parts processed in autoclaves are covered with a vacuum bag that is used primarily for compaction of laminates and to provide a path for removal of volatiles. The vacuum bag is also used to apply varying levels of vacuum to the part. Figure 13-10 shows an autoclave.

Heat bonder

A heat bonder, as shown in Fig. 13-11, is a portable device that automatically controls heating blankets based on temperature feedback from the repair area. Heat bonders also have a vacuum pump that supplies and monitors the vacuum in the vacuum bag. The heat bonder controls the cure cycle with thermocouples that are placed near the repair.

Types of Layups for Repair

When a composite aircraft is damaged and needs repair, a wet layup or prepreg layup can be used to repair the aircraft. Repairs should always be made per maintenance instruction manual.

Figure 13-10 Autoclave for processing of aerospace composite parts.

Figure 13-11   Portable heat bonder for composite repair.

Wet layup

During the wet layup process, a dry fabric is impregnated with a resin. Mix the resin system just before making the repair. Lay out the repair plies on a piece of fabric and impregnate the fabric with the resin. After the fabric is impregnated, cut the repair plies, stack in the correct ply orientation, and vacuum bag. Wet layup repairs are often used with fiberglass for nonstructural applications. Carbon and Kevlar® dry fabric could also be used with a wet layup resin system. Many resin systems used with wet layup cure at room temperature, are easy to accomplish, and the materials can be stored at room temperature for long period of times. The disadvantage of room temperature wet layup is that it does not restore the strength and durability of the original structure and parts that were cured at 250°F or 350°F during manufacturing. Some wet layup resins use an elevated temperature cure and have improved properties. In general, wet layup properties are less than properties of prepreg material.

Prepreg layup

Prepreg is a fabric or tape that is impregnated with a resin during the manufacturing process. The resin system is already mixed and is in the B-stage cure. Store the prepregs material in a freezer below 0°F to prevent further curing of the resin. The material is typically placed on a roll and a backing material is placed on one side of the material so that the prepreg does not stick together. The prepreg material is sticky and adheres to other plies easily during the stack-up process. You must remove the prepreg from the freezer and let the material thaw, which might take 8 hours for a full roll. Store the prepreg materials in a sealed, moisture proof bag. Do not open these bags until the material is completely thawed, to prevent contamination of the material by moisture. After the material is thawed and removed from the backing material, cut it in repair plies, stack in the correct ply orientation, and vacuum bag. Do not forget to remove the backing material when stacking the plies. Cure prepregs at an elevated cure cycle; the most common temperatures used are 250°F and 350°F. Autoclaves, curing ovens, and heat bonders can be used to cure the prepreg material.

Uncured prepreg materials have time limits for storage and use, as shown in Fig. 13-12. The maximum time allowed for storing of a prepreg at low temperature is called the storage life, which is typically 6 months to a year. The maximum time allowed for material at room temperature before the material cures is called the mechanical life. The recommended time at room temperature to complete layup and compaction is called the handling life. The handling life is shorter than the mechanical life. The mechanical life is measured from the time the material is removed from the freezer until the time the material is returned to the freezer.

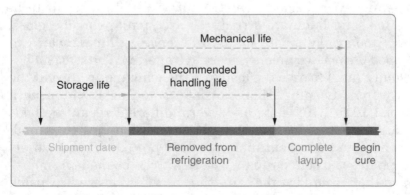

Figure 13-12 Prepeg time limits for storage and use.

Repairs of Honeycomb and Laminate Structures

This section will show typical repair techniques for honeycomb sandwich and laminate structures. Do not use this information to perform repairs but consult the applicable aircraft repair manual.

Honeycomb sandwich repair

Damage to face sheets due to impact damage or core damage due to moisture ingress are common for honeycomb sandwich structures and the part needs to be repaired or replaced to restore the integrity of the part. Debonding and delaminations are also common issues. The first step in the repair process is to inspect the damage to determine the size of the damage. Most types of damage can be detected with a simple tap test. Moisture ingress could be detected by a moisture detector, ultrasonic inspection, or x-ray. The second step is to remove the damaged material, as shown in Fig. 13-13. After the damage is removed a new core plug needs to be installed and the repair plies need to be cut and placed on the repair area as shown in Fig. 13-14. The next step would be to vacuum bag and cure the repair; see Fig. 13-15. After the cure is complete the repair needs to be inspected to determine if it is good.

Repair of laminate structure

Solid laminate structures have many more plies than the face sheets of honeycomb sandwich structures. The flush repair techniques for solid laminate structures are similar for fiberglass, Kevlar®, and graphite with minor differences. A flush repair can be stepped or, more commonly, scarved (tapered). The scarf angles are usually small to ease the load into the joint and to prevent the adhesive from escaping. This translates into thickness-to-length ratios of 1:10 to 1:70. There are several different repair methods for solid laminates. The patch can be precured and then secondarily bonded to the parent material. This procedure most closely approximates the bolted repair. The patch can be made from prepreg and then cocured at the same time as the adhesive. The patch can also be made using a wet layup repair. The curing cycle can

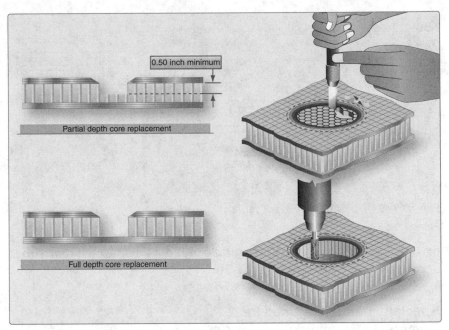

Figure 13-13 Honeycomb sandwich core removal.

also vary in length of time, cure temperature, and cure pressure, increasing the number of possible repair combinations.

Thick laminate structures can also be effectively repaired with a bolted repair similar to the repair of a metal aircraft. Titanium

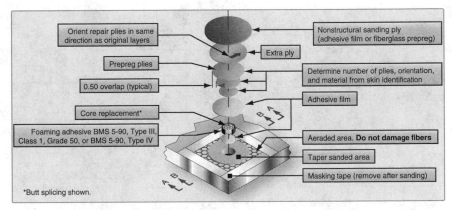

Figure 13-14 Installation of core and repair plies.

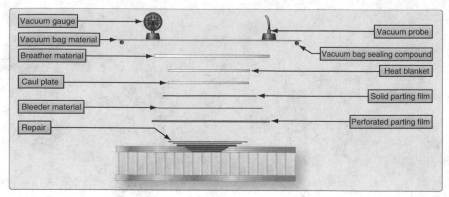

Figure 13-15   Vacuum bagging of repair.

or carbon fiber doublers are often used to repair carbon fiber structures as shown in Fig. 13-16.

Specialty Fasteners Used for Composite Structures

Many companies make specialty fasteners for composite structures and several types of fasteners are commonly used: threaded fasteners, lock bolts, blind bolts, blind rivets, and specialty fasteners for soft structures, such as honeycomb panels. The main differences between fasteners for metal and composite structures are the materials and the footprint diameter of nuts and collars.

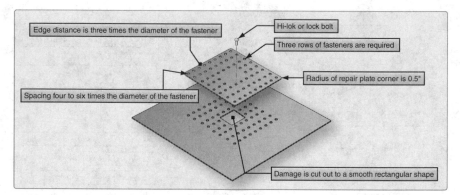

Figure 13-16   Bolted repair illustration.

Fastener Materials

Titanium alloy Ti-6Al-4V is the most common alloy for fasteners used with carbon fiber reinforced composite structures. Austenitic stainless steels, superalloys (e.g., A286), multiphase alloys (e.g., MP35N or MP159), and nickel alloys (e.g., alloy 718) also appear to be very compatible with carbon fiber composites.

Drilling

Hole drilling in composite materials is different from drilling holes in metal aircraft structures. Different types of drill bits, higher speeds, and lower feeds are required to drill precision holes. Structures made from carbon fiber and epoxy resin are very hard and abrasive, requiring special flat flute drills or similar four-flute drills. Aramid fiber (Kevlar®)/epoxy composites are not as hard as carbon but are difficult to drill unless special cutters are used because the fibers tend to fray or shred unless they are cut clean while embedded in the epoxy. Special drill bits with clothes pin points and fish-tail points have been developed that slice the fibers prior to pulling them out of the drilled hole. If the Kevlar®/epoxy part is sandwiched between two metal parts, standard twist drills can be used.

Drill bits used for carbon fiber and fiberglass are made from diamond-coated material or solid carbide because the fibers are so hard that standard high-speed steel (HSS) drill bits do not last long. Typically, twist drills are used, but brad point drills are also available. The Kevlar® fibers are not as hard as carbon, and standard HSS drill bits can be used. The hole quality can be poor if standard drill bits are used and the preferred drill style is the sickle-shaped Klenk drill (Fig. 13-17). This drill first pulls on the fibers and then shears them, which results in a better quality hole. Larger holes can be cut with diamond-coated hole saws.

Figure 13-17 Sickle-shaped Klenk drill.

Countersinking

Countersinking, a composite structure, is required when flush head fasteners are to be installed in the assembly. For metallic structures, a 100-degree included angle shear or tension head fastener has been the typical approach. In composite structures, two types of fastener are commonly used: a 100-degree included angle tension head fastener or a 130-degree included angle head fastener. Carbide cutters are used for producing a countersink in carbon/epoxy structure. These countersink cutters usually have straight flutes similar to those used on metals. For Kevlar® fiber/epoxy composites, S-shaped positive rake cutting flutes are used.

14

Standard Parts

Standard Parts Identification

Because the manufacture of aircraft requires a large number of miscellaneous small fasteners and other items usually called *hardware*, some degree of standardization is required. These standards have been derived by the various military organizations and described in detail in a set of specifications with applicable identification codes. These military standards have been universally adopted by the civil aircraft industry.

The derivation of a uniform standard is, by necessity, an evolutionary process. Originally, each of the military services derived its own standards. The old Army Air Corps set up AC (Air Corps) standards, whereas the Navy used NAF (Naval Aircraft Factory) standards. In time, these were consolidated into AN (Air Force-Navy) standards and NAS (National Aerospace Standards). Still later, these were consolidated into MS (Military Standard) designations.

At present, the three most common standards are:

- AN, Air Force-Navy.
- MS, Military Standard.
- NAS, National Aerospace Standards.

The aircraft mechanic will also occasionally be confronted with the following standard parts on older aircraft:

- AC (Air Corps).
- NAF (Naval Aircraft Factory).

Each of these standard parts is identified by its specification number and various dash numbers and letters to fully describe its name, size, and material.

Most air-frame manufacturers have need for special small parts and use their own series of numbers and specifications. However, they use the universal standard parts wherever practicable.

Because the purpose of this book is to provide the mechanic with a handy reference, only the most common standard parts are mentioned here with sufficient information to identify them.

More complete information on standard hardware is available from catalogs provided by the many aircraft parts suppliers.

Standard Parts Illustrations

AN standard parts, along with their equivalent and/or superseding MS numbers, are shown in the following pages.

AN Guide

AN 3 thru AN 20 BOLT — HEX HD. AIRCRAFT
AN 21 thru AN 36 BOLT — CLEVIS
AN 42 thru AN 49 BOLT — EYE
AN 73 thru AN 81 BOLT — DR HD (Engine)
AN 100 THIMBLE — CABLE
AN 115 SHACKLE — CABLE
AN 116 SHACKLE — SCREW PIN
AN 155 BARREL — TURNBUCKLE
AN 161 FORK — TURNBUCKLE
AN 162 FORK — TURNBUCKLE (For Bearing)
AN 165 EYE — TURNBUCKLE (For Pin)
AN 170 EYE — TURNBUCKLE (For cable)
AN 173 thru AN 186 BOLT, CLOSE TOL.
AN 210 thru AN 221 PULLEY — CONTROL
AN 253 PIN — HINGE
AN 254 SCREW — THUMB, NECKED

AN 255 SCREW — NECKED
AN 256 NUT — SELF LOCK (Rt. Angle Plate)
AN 257 HINGE — CONTINUOUS
AN 276 JOINT — BALL & SOCKET
AN 280 KEY — WOODRUFF
AN 295 CUP — OIL
AN 310 NUT — CASTLE (Air Frame)
AN 315 NUT — PLAIN (Air Frame)
AN 316 NUT — CHECK
AN 320 NUT — CASTLE, SHEAR
AN 335 NUT — PL. HEX (NC) (Semi-Fin)
AN 340 NUT — HEX, MACH. SCREW (NC)
AN 341 NUT — HEX, BRASS (Elec.)
AN 345 NUT — HEX, MACH. SCREW (NF)
AN 350 NUT — WING
AN 355 NUT — SLOTTED (Engine)
USAF 356 NUT — PAL

AN 360 NUT — PLAIN (Engine)

AN 362 NUT — PLATE, SELF-LOCK. (Hi-Temp.)

AN 363 NUT — HEX, SELF-LOCK. (Hi-Temp.)

AN 364 NUT — HEX, SELF-LOCK. (Thin)

AN 365 NUT — HEX, SELF-LOCK.

AN 366 NUT — PLATE, SELF-LOCK.

AN 373 NUT — PLATE, SELF-LOCK. (100° CTSK)

AN 380 PIN — COTTER

AN 381 PIN — COTTER, STAINLESS

AN 385 PIN — TAPERED, PLAIN

AN 386 PIN — THREADED TAPER

AN 392 thru AN 406 PIN — CLEVIS

AN 415 PIN — LOCK

AN 416 PIN — RETAINING, SAFETY

AN 426 RIVET — 100° FL. HD., ALUM.

AN 427 RIVET — 100° FL. HD., Steel, Monel, & Copper

AN 430 RIVET — RD. HD., ALUM.

AN 435 RIVET — RD. HD., Steel, Monel, & Copper

AN 442 RIVET — FL. HD., ALUM.

AN 450 RIVET — TUBULAR

AN 470 RIVET — UNIVERSAL HD., ALUM.

AN 481 CLEVIS — ROD END

AN 486 CLEVIS — ROD END ADJ.

AN 490 ROD END — THREADED

AN 500 SCREW — FILL. HD. (NC)

AN 501 SCREW — FILL. HD. (NF)

AN 502 SCREW — DR. FILL. HD. (Alloy Stl.) (NF)

AN 503 SCREW — DR. FILL. HD. (Alloy Stl.) (NC)

AN 504 SCREW — RD. HD. SELF TAP.

AN 505 SCREW — FLAT HD., 82° (NC)

AN 506 SCREW — FLAT HD., 82° SELF TAP.

AN 507 SCREW — FLAT HD., 100° (NF & NC)

AN 508 SCREW — RD. HD. BRASS (Elec.)

AN 509 SCREW — FL. HD. 100° (Structural) (ALLOY STEEL)

AN 510 SCREW — FLAT HD. 82° (NF)

AN 515 SCREW — RD. HD. (NC)

AN 520 SCREW — RD. HD. (NF)

AN 525 SCREW — WASHER HD. (Alloy Stl.)

AN 526 SCREW — TRUSS HD. (NF & NC)

AN 530 SCREW — RD. HD., SHEET METAL (TYPE B)

AN 531 SCREW — FL. HD., 82° SHEET METAL (TYPE B)

AN 535 SCREW — RD. HD. DRIVE (Type "U")

AN 545 SCREW — WOOD, RD. HD.

AN 550 SCREW — WOOD, FLAT HD.

AN 565 SCREW — HDLESS., SET

AN 663 TERMINAL — CABLE, DBLD. SHK. BALL (FOR SWAGING)

AN 664 TERMINAL — CABLE, SGLE. SHK. BALL (FOR SWAGING)

AN 665 TERMINAL — CABLE, THDED. CLEVIS

AN 666 TERMINAL — CABLE, THDED. (FOR SWAGING)

AN 667 TERMINAL — CABLE, FORK END (FOR SWAGING)

AN 668 TERMINAL — CABLE, EYE END (FOR SWAGING)

AN 669 TERMINAL — CABLE, TURNBUCKLE (FOR SWAGING)

AN 737 CLAMP — HOSE

AN 741 CLAMP — TUBE

AN 742 CLAMP — PLAIN, SUPPORT

AN 900 GASKET — COP. — ASBESTOS, ANGULAR

AN 901 GASKET — METAL TUBE

AN 931 GROMMET — ELASTIC

AN 935 WASHER — LOCK, SPRING

AN 936 WASHER — LOCK TOOTH (Ext. & Int.)

AN 960 WASHER — FLAT, AIRCRAFT

AN 961 WASHER — FLAT, BRASS (Elec.)

AN 970 WASHER — FLAT, LARGE AREA

AN 975 WASHER — TAPER PIN

AN 996 RING — LOCK

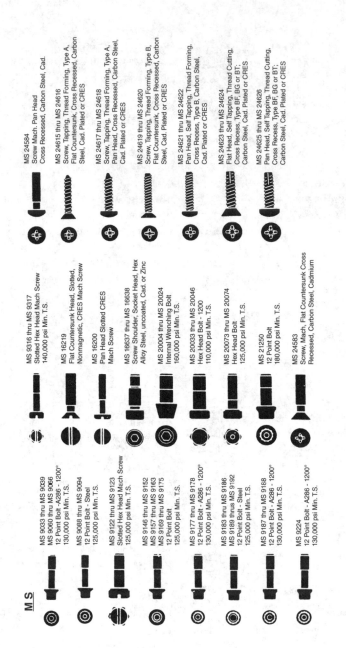

MS 9033 thru MS 9039
MS 9060 thru MS 9066
12 Point Bolt - A286 - 1200°
130,000 psi Min. T.S.

MS 9088 thru MS 9094
12 Point Bolt - Steel
125,000 psi Min.T.S.

MS 9122 thru MS 9123
Slotted Hex Head Mach Screw
125,000 psi Min. T.S.

MS 9146 thru MS 9152
MS 9157 thru MS 9163
MS 9169 thru MS 9175
12 Point Bolt
125,000 psi Min. T.S.

MS 9177 thru MS 9178
12 Point Bolt - A286 - 1200°
130,000 psi Min. T.S.

MS 9183 thru MS 9186
MS 9189 thru MS 9192
12 Point Bolt - Steel
125,000 psi Min. T.S.

MS 9187 thru MS 9188
12 Point Bolt - A286 - 1200°
130,000 psi Min. T.S.

MS 9224
12 Point Bolt - A286 - 1200°
130,000 psi Min. T.S.

MS 9316 thru MS 9317
Slotted Hex Head Mach Screw
140,000 psi Min. T.S.

MS 16219
Flat Countersunk Head, Slotted,
Nonmagnetic, CRES Mach Screw

MS 16200
Pan Head Slotted CRES
Mach Screw

MS 16637 thru MS 16638
Screw Shoulder, Socket Head, Hex
Alloy Steel, uncoated, Cad. or Zinc

MS 20004 thru MS 20024
Internal Wrenching Bolt
160,000 psi Min. T.S.

MS 20033 thru MS 20046
Hex Head Bolt - 1200
110,000 psi Min. T.S.

MS 20073 thru MS 20074
Hex Head Bolt
125,000 psi Min. T.S.

MS 21250
12 Point Bolt
180,000 psi Min. T.S.

MS 24583
Screw, Mach, Flat Countersunk Cross
Recessed, Carbon Steel, Cadmium

MS 24584
Screw Mach. Pan Head
Cross Recessed, Carbon Steel, Cad.

MS 24615 thru MS 24616
Screw, Tapping, Thread Forming, Type A,
Flat Countersunk, Cross Recessed, Carbon
Steel, Cad. Plated or CRES

MS 24617 thru MS 24618
Screw, Tapping, Thread Forming, Type A,
Pan Head, Cross Recessed, Carbon Steel,
Cad. Plated or CRES

MS 24619 thru MS 24620
Screw, Tapping, Thread Forming, Type B,
Flat Countersunk, Cross Recessed, Carbon
Steel, Cad. Plated or CRES

MS 24621 thru MS 24622
Pan Head, Self Tapping, Thread Forming,
Cross Recess, Type B, Carbon Steel,
Cad. Plated or CRES

MS 24623 thru MS 24624
Flat Head, Self Tapping, Thread Cutting,
Cross Recess, Type BF, BG or BT;
Carbon Steel, Cad. Plated or CRES

MS 24625 thru MS 24626
Pan Head, Self Tapping, Thread Cutting,
Cross Recess, Type BF, BG or BT;
Carbon Steel, Cad. Plated or CRES

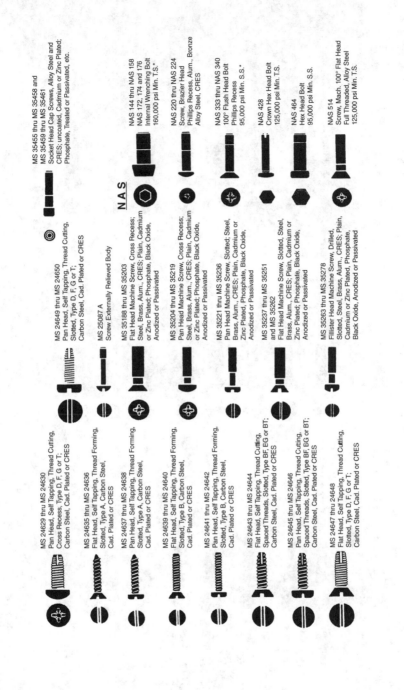

MS 35455 thru MS 35458 and
MS 35459 thru MS 35461
Socket Head Cap Screws, Alloy Steel and
CRES; uncoated, Cadmium or Zinc Plated;
Phosphate, Treated or Passivated, etc.

MS 24649 thru MS 24650
Pan Head, Self Tapping, Thread Cutting,
Slotted, Type D, F, G or T;
Carbon Steel, Cad. Plated or CRES

MS 25087 A
Screw Externally Relieved Body

MS 35188 thru MS 35203
Flat Head Machine Screw, Cross Recess;
Steel, Brass, Alum., CRES; Plain, Cadmium
or Zinc Plated; Phosphate, Black Oxide,
Anodized or Passivated

MS 35204 thru MS 35219
Pan Head Machine Screw, Cross Recess;
Steel, Brass, Alum., CRES; Plain, Cadmium
or Zinc Plated; Phosphate, Black Oxide,
Anodized or Passivated

MS 35221 thru MS 35236
Pan Head Machine Screw, Slotted; Steel,
Brass, Alum., CRES; Plain, Cadmium or
Zinc Plated, Phosphate, Black Oxide,
Anodized or Passivated

MS 35237 thru MS 35251
and MS 35262
Flat Head Machine Screw, Slotted, Steel,
Brass, Alum., CRES; Plain, Cadmium or
Zinc Plated; Phosphate, Black Oxide,
Anodized or Passivated

MS 35263 thru MS 35278
Fillister Head Machine Screw, Drilled,
Slotted, Steel, Brass, Alum., CRES; Plain,
Cadmium or Zinc Plated, Phosphate,
Black Oxide, Anodized or Passivated

N A S

NAS 144 thru NAS 158
NAS 172, 174 and 176
Internal Wrenching Bolt
160,000 psi Min. T.S.*

NAS 220 thru NAS 224
Screw, Brazier Head
Phillips Recess, Alum., Bronze
Alloy Steel, CRES

NAS 333 thru NAS 340
100° Flush Head Bolt
Phillips Recess
95,000 psi Min. S.S.*

NAS 428
Crown Hex Head Bolt
125,000 psi Min. T.S.

NAS 464
Hex Head Bolt
95,000 psi Min. S.S.

NAS 514
Screw, Mach. 100° Flat Head
Full Threaded, Alloy Steel
125,000 psi Min. T.S.

MS 24629 thru MS 24630
Pan Head, Self Tapping, Thread Cutting,
Cross Recess, Type D, F, G or T;
Carbon Steel, Cad. Plated or CRES

MS 24635 thru MS 24636
Flat Head, Self Tapping, Thread Forming,
Slotted, Type A, Carbon Steel,
Cad. Plated or CRES

MS 24637 thru MS 24638
Pan Head, Self Tapping, Thread Forming,
Slotted, Type A, Carbon Steel,
Cad. Plated or CRES

MS 24639 thru MS 24640
Flat Head, Self Tapping, Thread Forming,
Slotted, Type B, Carbon Steel,
Cad. Plated or CRES

MS 24641 thru MS 24642
Pan Head, Self Tapping, Thread Forming,
Slotted, Type B, Carbon Steel,
Cad. Plated or CRES

MS 24643 thru MS 24644
Flat Head, Self Tapping, Thread Cutting,
Spaced Threads, Slotted, Type BF, EG or BT;
Carbon Steel, Cad. Plated or CRES

MS 24645 thru MS 24646
Pan Head, Self Tapping, Thread Cutting,
Spaced Threads, Slotted, Type BF, BG or BT;
Carbon Steel, Cad. Plated or CRES

MS 24647 thru 24648
Flat Head, Self Tapping, Thread Cutting,
Slotted, Type D, F, G or T;
Carbon Steel, Cad. Plated or CRES

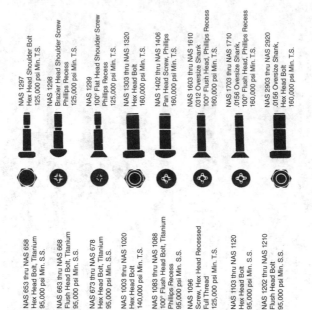

NAS 1297
Hex Head Shoulder Bolt
125,000 psi Min. T.S.

NAS 1298
Brazier Head Shoulder Screw
Phillips Recess
125,000 psi Min. T.S.

NAS 1299
100° Flat Head Shoulder Screw
Phillips Recess
125,000 psi Min. T.S.

NAS 1303 thru NAS 1320
Hex Head Bolt
160,000 psi Min. T.S.

NAS 1402 thru NAS 1406
Pan Head Screw, Phillips
160,000 psi Min. T.S.

NAS 1603 thru NAS 1610
0312 Oversize Shank
100° Flush Head, Phillips Recess
160,000 psi Min. T.S.

NAS 1703 thru NAS 1710
0156 Oversize Shank,
100° Flush Head, Phillips Recess
160,000 psi Min. T.S.

NAS 2903 thru NAS 2920
0156 Oversize Shank,
Hex Head Bolt
160,000 psi Min. T.S.

NAS 653 thru NAS 658
Hex Head Bolt, Titanium
95,000 psi Min. S.S.

NAS 663 thru NAS 668
Flush Head Bolt, Titanium
95,000 psi Min. S.S.

NAS 673 thru NAS 678
Hex Head Bolt, Titanium
95,000 psi Min. S.S.

NAS 1003 thru NAS 1020
Hex Head Bolt
140,000 psi Min. T.S.

NAS 1083 thru NAS 1088
100° Flush Head Bolt, Titanium
Phillips Recess
95,000 psi Min. S.S.

NAS 1096
Screw, Hex Head Recessed
Full Thread
125,000 psi Min. T.S.

NAS 1103 thru NAS 1120
Hex Head Bolt
95,000 psi Min. S.S.

NAS 1202 thru NAS 1210
Flush Head Bolt
95,000 psi Min. S.S.

NAS 517
100° Flush Head Bolt
95,000 psi Min. S.S.

NAS 560
Screw - Hi Temp
100° Flush Head
321, A286 or Inconel "X"

NAS 563–572
Hex Head Bolt
160,000 psi Min. T.S.

NAS 600 thru NAS 606
Screw, Mach. Pan Head, Phillip
Full Threaded, Alloy Steel
160,000 psi Min. T.S.

NAS 608 - NAS 609
Std. Socket Head Cap Screw

NAS 610 thru NAS 616
Pan Head Screw
Reed & Prince Recess
160,000 psi Min. T.S.

NAS 623
Pan Head Screw
Phillips Recess
160,000 psi Min. T.S.

NAS 624 thru NAS 644
12 Point Bolt
180,000 psi Min. T.S.

AN3–AN20 General-Purpose Bolt

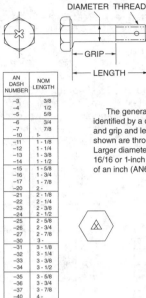

DIAMETER THREAD

←— GRIP —→

←——— LENGTH ———→

AN DASH NUMBER	NOM LENGTH
−3	3/8
−4	1/2
−5	5/8
−6	3/4
−7	7/8
−10	1-
−11	1 - 1/8
−12	1 - 1/4
−13	1 - 3/8
−14	1 - 1/2
−15	1 - 5/8
−16	1 - 3/4
−17	1 - 7/8
−20	2 -
−21	2 - 1/8
−22	2 - 1/4
−23	2 - 3/8
−24	2 - 1/2
−25	2 - 5/8
−26	2 - 3/4
−27	2 - 7/8
−30	3 -
−31	3 - 1/8
−32	3 - 1/4
−33	3 - 3/8
−34	3 - 1/2
−35	3 - 5/8
−36	3 - 3/4
−37	3 - 7/8
−40	4 -

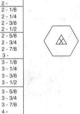

NON-CORROSIVE-RESISTANT STEEL MACHINE BOLTS SHOWN, MEET SPECIFICATION MIL-B-6812. CADMIUM PLATED TO SPECIFICATION QQ-P-416. DRILLED HEAD BOLTS ARE COUNTERSINK DRILLED.

PART NUMBER EXAMPLES FOR A CADMIUM PLATED STEEL BOLT HAVING A DIAMETER OF 3/8" A NOMINAL LENGTH OF 1":
AN6-10 (DRILLED SHANK)
AN6H10 (DRILLED HEAD AND SHANK)
AN6-10A (UNDRILLED)
AN6H10A (DRILLED HEAD)

The general-purpose structural bolt (AN3 through AN20) is identified by a cross or asterisk. Nominal lengths are shown above and grip and length and tolerances are shown below. Examples shown are through AN8 (1/2 inch) and lengths through -40 (4 inch). Larger diameters are identified by sixteenths of an inch (AN16, 16/16 or 1-inch diameter). Lengths are correspondingly coded in 8ths of an inch (AN63 = 6 inch + 3/8 inch or 6 3/8 inch).

AN173-AN176 CLOSE TOLERANCE BOLTS

AN173 thru AN186 bolts are cadmium-plated steel with shanks drilled or undrilled and heads drilled or undrilled.
AN175-10 (drill shank only) (5/16 inch diameter)
AN175-10A (undrilled shank, undrilled head)
AN175-H10A (drilled head, undrilled shank)
AN175-H10 (drilled head, drilled shank)
Dimensions and coding similar to AN3-AN20 bolts. The third number indicated the bolt diameter in sixteenths.

DASH NO.	AN3 GRIP ±1/64	AN3 LENGTH +1/32 −1/64	AN4 GRIP ±1/64	AN4 LENGTH +1/32 −1/64	AN5 GRIP ±1/64	AN5 LENGTH +1/32 −1/64	AN6 GRIP ±1/64	AN6 LENGTH +1/32 −1/64	AN7 GRIP ±1/64	AN7 LENGTH +1/32 −1/64	AN8 GRIP ±1/64	AN8 LENGTH +1/32 −1/64
3	1/16	15/32	1/16	15/32								
4	1/8	17/32	1/16	17/32	1/16	19/32						
5	1/4	21/32	3/16	21/32	3/16	23/32	1/16	45/64	1/16	23/32		
6	3/8	25/32	5/16	25/32	5/16	27/32	3/16	53/64	3/16	27/32	1/16	27/32
7	1/2	29/32	7/16	29/32	7/16	31/32	5/16	61/64	5/16	31/32	3/16	31/32
10	5/8	1 - 1/32	9/16	1 - 1/32	9/16	1 - 3/32	7/16	1 - 5/64	7/16	1 - 3/32	5/16	1 - 3/32
11	3/4	1 - 5/32	11/16	1 - 5/32	11/16	1 - 7/32	9/16	1 - 13/64	9/16	1 - 7/32	7/16	1 - 7/32
12	7/8	1 - 9/32	13/16	1 - 9/32	13/16	1 - 11/32	11/16	1 - 21/64	11/16	1 - 11/32	9/16	1 - 11/32
13	1	1 -13/32	15/16	1 -13/32	15/16	1 - 15/32	13/16	1 - 29/64	13/16	1 - 15/32	11/16	1 - 15/32
14	1 - 1/8	1 - 17/32	1 - 1/16	1 - 17/32	1 - 1/16	1 - 19/32	15/16	1 - 37/64	15/16	1 - 19/32	13/16	1 - 19/32
15	1 - 1/4	1 - 21/32	1 - 3/16	1 - 21/32	1 - 3/16	1 - 23/32	1 - 1/16	1 - 45/64	1 - 1/16	1 - 23/32	15/16	1 - 23/32
16	1 - 3/8	1 -25/32	1 - 5/16	1 -25/32	1 - 5/16	1 - 27/32	1 - 3/16	1 - 53/64	1 - 3/16	1 - 27/32	1 - 1/16	1 - 27/32
17	1 - 1/2	1 -29/32	1 - 7/16	1 -29/32	1 - 7/16	1 - 31/32	1 - 5/16	1 - 61/64	1 - 5/16	1 - 31/32	1 - 3/16	1 - 31/32
20	1 - 5/8	2 - 1/32	1 - 9/16	2- 1/32	1 - 9/16	2- 3/32	1 - 7/16	2 - 5/64	1 - 7/16	2 - 3/32	1 - 5/16	2- 3/32
21	1 - 3/4	2 - 5/32	1 - 11/16	2- 5/32	1 - 11/16	2- 7/32	1 - 9/16	2 - 13/64	1 - 9/16	2 - 7/32	1 - 7/16	2- 7/32
22	1 - 7/8	2 - 9/32	1 - 13/16	2- 9/32	1 - 13/16	2- 11/32	1 - 11/16	2 - 21/64	1 - 11/16	2 - 11/32	1 - 9/16	2- 11/32
23	2	2 -13/32	1 - 15/16	2-13/32	1 - 15/16	2- 15/32	1 - 13/16	2 - 29/64	1 - 13/16	2 - 15/32	1 - 11/16	2- 15/32
24	2 - 1/8	2 -17/32	2 - 1/16	2-17/32	2 - 1/16	2- 19/32	1 - 15/16	2 - 37/64	1 - 15/16	2 - 19/32	1 - 13/16	2- 19/32
25	2 - 1/4	2 -21/32	2 - 3/16	2-21/32	2 - 3/16	2- 23/32	2 - 1/16	2 - 45/64	2 - 1/16	2 - 23/32	1 - 15/16	2- 23/32
26	2 - 3/8	2 -25/32	2 - 5/16	2-25/32	2 - 5/16	2- 27/32	2 - 3/16	2 - 53/64	2 - 3/16	2 - 27/32	2 - 1/16	2- 27/32
27	2 - 1/2	2 -29/32	2 - 7/16	2-29/32	2 - 7/16	2- 31/32	2 - 5/16	2 - 61/64	2 - 5/16	2 - 31/32	2 - 3/16	2- 31/32
30	2 - 5/8	3 - 1/32	2 - 9/16	3- 1/32	2 - 9/16	3- 3/32	2 - 7/16	3 - 5/64	2 - 7/16	3 - 3/32	2 - 5/16	3- 3/32
31	2 - 3/4	3 - 5/32	2 - 11/16	3- 5/32	2 - 11/16	3- 7/32	2 - 9/16	3 - 13/64	2 - 9/16	3 - 7/32	2 - 7/16	3- 7/32
32	2 - 7/8	3 - 9/32	2 - 13/16	3- 9/32	2 - 13/16	3- 11/32	2 - 11/16	3 - 21/64	2 - 11/16	3 - 11/32	2 - 9/16	3- 11/32
33	3	3 -13/32	2 - 15/16	3-13/32	2 - 15/16	3- 15/32	2 - 13/16	3 - 29/64	2 - 13/16	3 - 15/32	2 - 11/16	3- 15/32
34	3 - 1/8	3 -17/32	3 - 1/16	3- 17/32	3 - 1/16	3- 19/32	2 - 15/16	3 - 37/64	2 - 15/16	3 - 19/32	2 - 13/16	3- 19/32
35	3 - 1/4	3 -21/32	3 - 3/16	3- 21/32	3 - 3/16	3- 23/32	3 - 1/16	3 - 45/64	3 - 1/16	3 - 23/32	2 - 15/16	3- 23/32
36	3 - 3/8	3 -25/32	3 - 5/16	3- 25/32	3 - 5/16	3- 27/32	3 - 3/16	3 - 53/64	3 - 3/16	3 - 27/32	3 - 1/16	3- 27/32
37	3 - 1/2	3 -29/32	3 - 7/16	3- 29/32	3 - 7/16	3- 31/32	3 - 5/16	3 - 61/64	3 - 5/16	3 - 31/32	3 - 3/16	3- 31/32
40	3 - 5/8	4 - 1/32	3 - 9/16	4 - 1/32	3 - 9/16	4- 3/32	3 - 7/16	4 - 5/64	3 - 7/16	4 - 3/32	3 - 5/16	4- 3/32
41	3 - 3/4	4 - 5/32	3 - 11/16	4- 5/32	3 - 11/16	4- 7/32	3 - 9/16	4 - 13/64	3 - 9/16	4 - 7/32	3 - 7/16	4- 7/32
42	3 - 7/8	4 - 9/32	3 - 13/16	4- 9/32	3 - 13/16	4- 11/32	3 - 11/16	4 - 21/64	3 - 11/16	4 - 11/32	3 - 9/16	4- 11/32
43	4	4 -13/32	3 - 15/16	4-13/32	3 - 15/16	4- 15/32	3 - 13/16	4 - 29/64	3 - 13/16	4 - 15/32	3 - 11/16	4- 15/32
44	4 - 1/8	4 -17/32	4 - 1/16	4-17/32	4 - 1/16	4- 19/32	3 - 15/16	4 - 37/64	3 - 15/16	4 - 19/32	3 - 13/16	4- 19/32
45	4 - 1/4	4 -21/32	4 - 3/16	4-21/32	4 - 3/16	4- 23/32	4 - 1/16	4 - 45/64	4 - 1/16	4 - 23/32	3 - 15/16	4- 23/32
46	4 - 3/8	4 -25/32	4 - 5/16	4-25/32	4 - 5/16	4- 27/32	4 - 3/16	4 - 53/64	4 - 3/16	4 - 27/32	4 - 1/16	4- 27/32
47	4 - 1/2	4 -29/32	4 - 7/16	4-29/32	4 - 7/16	4- 31/32	4 - 5/16	4 - 61/64	4 - 5/16	4 - 31/32	4 - 3/16	4- 31/32
50	4 - 5/8	5 - 1/32	4 - 9/16	5- 1/32	4 - 9/16	5- 3/32	4 - 7/16	5 - 5/64	4 - 7/16	5 - 3/32	4 - 5/16	5- 3/32
51	4 - 3/4	5 - 5/32	4 - 11/16	5- 5/32	4 - 11/16	5- 7/32	4 - 9/16	5 - 13/64	4 - 9/16	5 - 7/32	4 - 7/16	5- 7/32
52	4 - 7/8	5 - 9/32	4 - 13/16	5- 9/32	4 - 13/16	5- 11/32	4 - 11/16	5 - 21/64	4 - 11/16	5 - 11/32	4 - 9/16	5- 11/32
53	5	5 -13/32	4 - 15/16	5-13/32	4 - 15/16	5- 15/32	4 - 13/16	5 - 29/64	4 - 13/16	5 - 15/32	4 - 11/16	5- 15/32
54	5 - 1/8	5 -17/32	5 - 1/16	5- 17/32	5 - 1/16	5- 19/32	4 - 15/16	5 - 37/64	4 - 15/16	5 - 19/32	4 - 13/16	5- 19/32
55	5 - 1/4	5 -21/32	5 - 3/16	5- 21/32	5 - 3/16	5- 23/32	5 - 1/16	5 - 45/64	5 - 1/16	5 - 23/32	4 - 15/16	5- 23/32
56	5 - 3/8	5 -25/32	5 - 5/16	5- 25/32	5 - 5/16	5- 27/32	5 - 3/16	5 - 53/64	5 - 3/16	5 - 27/32	5 - 1/16	5- 27/32
57	5 - 1/2	5 -29/32	5 - 7/16	5-29/32	5 - 7/16	5- 31/32	6	5 - 61/64	5 - 5/16	5 - 31/32	5 - 3/16	5- 31/32
60	5 - 5/8	6 - 1/32	5 - 9/16	6- 1/32	5 - 9/16	6- 3/32	5 - 7/16	6 - 5/64	5 - 7/16	6 - 3/32	5 - 5/16	6- 3/32
61	5 - 3/4	6 - 5/32	5 - 11/16	6- 5/32	5 - 11/16	6- 7/32	5 - 9/16	6 - 13/64	5 - 9/16	6 - 7/32	5 - 7/16	6- 7/32
62	5 - 7/8	6 - 9/32	5 - 13/16	6- 9/32	5 - 13/16	6- 11/32	5 - 11/16	6 - 21/64	5 - 11/16	6 - 11/32	5 - 9/16	6- 11/32
63	6	6 -13/32	5 - 15/16	6-13/32	5 - 15/16	6- 15/32	5 - 13/16	6 - 29/64	5 - 13/16	6 - 15/32	5 - 11/16	6- 15/32
64	6 - 1/8	6 -17/32	6 - 1/16	6- 17/32	6 - 1/16	6- 19/32	5 - 15/16	6 - 37/64	5 - 15/16	6 - 19/32	5 - 13/16	6- 19/32
65	6 - 1/4	6 -21/32	6 - 3/16	6- 21/32	6 - 3/16	6- 23/32	6 - 1/16	6 - 45/64	6 - 1/16	6 - 23/32	5 - 15/16	6- 23/32
66	6 - 3/8	6 -25/32	6 - 5/16	6- 25/32	6 - 5/16	6- 27/32	6 - 3/16	6 - 53/64	6 - 3/16	6 - 27/32	6 - 1/16	6- 27/32
67	6 - 1/2	6 -29/32	6 - 7/16	6- 29/32	6 - 7/16	6- 31/32	6 - 5/16	6 - 61/64	6 - 5/16	6 - 31/32	6 - 3/16	6- 31/32
70	7 - 5/8	7 - 1/32	6 - 9/16	7- 1/32	7 - 1/32	6- 9/16	7 - 3/32	6 - 7/16	7 - 5/64	6 - 7/16	7 - 3/32	6 - 5/16

AN21 – AN36 Clevis Bolt

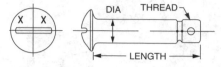

PART NUMBER EXAMPLES FOR CLEVIS BOLT HAVING A #10–32 DIAMETER AND NOMINAL LENGTH OF 15/16".

AN23–15   (HAS COTTER PIN HOLE)
AN23–15A   (NO COTTER PIN HOLE)

The clevis bolt is used for shear loads only and requires a shear nut AN320 (for use with cotter pin) or AN364 (MS 20364) self-locking nut. Nominal sizes, grip length, and tolerances are shown. Only AN23, 24, and 25 are shown. Other diameters are indicated by AN number. For example, AN28 is $8/16$ or $1/2$ inch diameter. Lengths are in sixteenths of an inch, –18 is 18 sixteenths or $1\,1/8$ inch long.

DASH NO.	AN23 GRIP	AN23 LENGTH	AN24 GRIP	AN24 LENGTH	AN25 GRIP	AN25 LENGTH
8	3/16	17/32	3/16	17/32		
9	1/4	19/32	1/4	19/32	1/4	39/64
10	5/16	21/32	5/16	21/32	5/16	43/64
11	3/8	23/32	3/8	23/32	3/8	47/64
12	7/16	25/32	7/16	25/32	7/16	51/64
13	1/2	27/32	1/2	27/32	1/2	55/64
14	9/16	29/32	9/16	29/32	9/16	59/64
15	5/8	31/32	5/8	31/32	5/8	63/64
16	11/16	1 - 1/32	11/16	1 - 1/32	11/16	1 - 3/64
17	3/4	1 - 3/32	3/4	1 - 3/32	3/4	1 - 7/64
18	13/16	1 - 5/32	13/16	1 - 5/32	13/16	1 - 11/64
19	7/8	1 - 7/32	7/8	1 - 7/32	7/8	1 - 15/64
20	15/16	1 - 9/32	15/16	1 - 9/32	15/16	1 - 19/64
21	1	1 - 11/32	1	1 - 11/32	1	1 - 23/64
22	1 - 1/16	1 - 13/32	1 - 1/16	1 - 13/32	1 - 1/16	1 - 27/64
23	1 - 1/8	1 - 15/32	1 - 1/8	1 - 15/32	1 - 1/8	1 - 31/64
24	1 - 3/16	1 - 17/32	1 - 3/16	1 - 17/32	1 - 3/16	1 - 35/64
25	1 - 1/4	1 - 19/32	1 - 1/4	1 - 19/32	1 - 1/4	1 - 39/64
26	1 - 5/16	1 - 21/32	1 - 5/16	1 - 21/32	1 - 5/16	1 - 43/64
27	1 - 3/8	1 - 23/32	1 - 3/8	1 - 23/32	1 - 3/8	1 - 47/64
28	1 - 7/16	1 - 25/32	1 - 7/16	1 - 25/32	1 - 7/16	1 - 51/64
29	1 - 1/2	1 - 27/32	1 - 1/2	1 - 27/32	1 - 1/2	1 - 55/64
30	1 - 9/16	1 - 29/32	1 - 9/16	1 - 29/32	1 - 9/16	1 - 59/64
31	1 - 5/8	1 - 31/32	1 - 5/8	1 - 31/32	1 - 5/8	1 - 63/64
32	1 - 11/16	2 - 1/32	1 - 11/16	2 - 1/32	1 - 11/16	2 - 3/64
34	1 - 13/16	2 - 5/32	1 - 13/16	2 - 5/32	1 - 13/16	2 - 11/64
36	1 - 15/16	2 - 9/32	1 - 15/16	2 - 9/32	1 - 15/16	2 - 19/64
38	2 - 1/16	2 - 13/32	2 - 1/16	2 - 13/32	2 - 1/16	2 - 27/64
40	2 - 3/16	2 - 17/32	2 - 3/16	2 - 17/32	2 - 3/16	2 - 35/64
42	2 - 5/16	2 - 21/32	2 - 5/16	2 - 21/32	2 - 5/16	2 - 43/64
44	2 - 7/16	2 - 25/32	2 - 7/16	2 - 25/32	2 - 7/16	2 - 51/64
46	2 - 9/16	2 - 29/32	2 - 9/16	2 - 29/32	2 - 9/16	2 - 59/64
48	2 - 11/16	3 - 1/32	2 - 11/16	3 - 1/32	2 - 11/16	3 - 3/64
50	2 - 13/16	3 - 5/32	2 - 13/16	3 - 5/32	2 - 13/16	3 - 11/64
52	2 - 15/16	3 - 9/32	2 - 15/16	3 - 9/32	2 - 15/16	3 - 19/64
54	3 - 1/16	3 - 13/32	3 - 1/16	3 - 13/32	3 - 1/16	3 - 27/64
56	3 - 3/16	3 - 17/32	3 - 3/16	3 - 17/32	3 - 3/16	3 - 35/64
58	3 - 5/16	3 - 21/32	3 - 5/16	3 - 21/32	3 - 5/16	3 - 43/64
60	3 - 7/16	3 - 25/32	3 - 7/16	3 - 25/32	3 - 7/16	3 - 51/64
62	3 - 9/16	3 - 29/32	3 - 9/16	3 - 29/32	3 - 9/16	3 - 59/64
64	3 - 11/16	4 - 1/32	3 - 11/16	4 - 1/32	3 - 11/16	4 - 3/64
66			3 - 13/16	4 - 5/32		
68			3 - 15/16	4 - 9/32		
70			4 - 1/16	4 - 13/32		
72			4 - 3/16	4 - 17/32		

	AN23	AN24	AN25
	#10	1/4	5/16
	–32	–28	–24

AN DASH NUMBER	NOMINAL LENGTH
–8	1/2
–9	9/16
–10	5/8
–11	11/16
–12	3/4
–13	13/16
–14	7/8
–15	15/16
–16	1
–17	1 - 1/16
–18	1 - 1/8
–19	1 - 3/16
–20	1 - 1/4
–21	1 - 5/16
–22	1 - 3/8
–23	1 - 7/16
–24	1 - 1/2
–25	1 - 9/16
–26	1 - 5/8
–27	1 - 11/16
–28	1 - 3/4
–29	1 - 13/16
–30	1 - 7/8
–31	1 - 15/16
–32	2

AN42–AN49 Eye Bolt

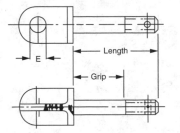

AN#	SIZE	E MIN	E MAX	PIN SIZE
AN42B	10–32	.190	.192	3/16
AN43B	1/4–28	.190	.192	3/16
AN44	5/16–24	.250	.253	1/4
AN45	5/16–24	.313	.316	5/16
AN46	3/8–24	.375	.378	3/8
AN47	7/16–20	.375	.378	3/8
AN48	1/2–20	.438	.441	7/16
AN49	9/16–18	.500	.503	1/2

Dash numbers for grip and length are the same as those for aircraft bolts AN3–AN20 of the same body diameter. Example: AN43-12 is eye bolt. 1/4 inch diameter, 3/16 eye and 1 1/4 inch long (add A for absence of hole).

AN392–AN406 (MS20392) Clevis Pin

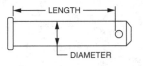

DIA.	AN PART NO.	MS 20392 BASIC NO.	DASH NO. RANGE
1/8	AN392	–1C	–7 thru –67
3/16	AN393	–2C	–7 thru –95
1/4	AN394	–3C	–11 thru –97
5/16	AN395	–4C	–11 thru –97
3/8	AN396	–5C	–15 thru –127
7/16	AN397	–6C	–15 thru –127
1/2	AN398	–7C	–15 thru –127
9/16	AN399	–8C	–15 thru –127
5/8	AN400	–9C	–15 thru –127
3/4	AN402	–10C	–15 thru –127
7/8	AN404	–11C	–19 thru –137
1"	AN406	–12C	–19 thru –137

Dash Numbers are the Grip Length as Expressed in ODD 1/32's of an Inch ONLY.

Example: AN395-41 is a 5/16 inch diameter pin with an effective length of 19/32 inch. Equivalent MS number is MS 20392-4C41.

1/8 DIAMETER			3/16 DIAMETER			1/4 DIAMETER		
AN392 DASH NO.	MS 20392 DASH NO.	LENGTH	AN393 DASH NO.	MS 20392 DASH NO.	LENGTH	AN394 DASH NO.	MS 20392 DASH NO.	LENGTH
–7	1C7	7/32	–7	2C7	7/32	–11	3C11	11/32
–9	1C9	9/32	–9	2C9	9/32	–13	3C13	13/32
–11	1C11	11/32	–11	2C11	11/32	–15	3C15	15/32
–13	1C13	13/32	–13	2C13	13/32	–17	3C17	17/32
–15	1C15	15/32	–15	2C15	15/32	–19	3C19	19/32
–17	1C17	17/32	–17	2C17	17/32	–21	3C21	21/32
–19	1C19	19/32	–19	2C19	19/32	–23	3C23	23/32
–21	1C21	21/32	–21	2C21	21/32	–25	3C25	25/32
–23	1C23	23/32	–23	2C23	23/32	–27	3C27	27/32
–25	1C25	25/32	–25	2C25	25/32	–29	3C29	29/32
–27	1C27	27/32	–27	2C27	27/32	–31	3C31	31/32
–29	1C29	29/32	–29	2C29	29/32	–33	3C33	1-1/32
–31	1C31	31/32	–31	2C31	31/32	–35	3C35	1-3/32
–33	1C33	1-1/32	–33	2C33	1-1/32	–37	3C37	1-5/32
–35	1C35	1-3/32	–35	2C35	1-3/32	–39	3C39	1-7/32
–37	1C37	1-5/32	–37	2C37	1-5/32	–41	3C41	1-9/32
–39	1C39	1-7/32	–39	2C39	1-7/32	–43	3C43	1-11/32
–41	1C41	1-9/32	–41	2C41	1-9/32	–45	3C45	1-13/32
–43	1C43	1-11/32	–43	2C43	1-11/32	–47	3C47	1-15/32
–45	1C45	1-13/32	–45	2C45	1-13/32	–49	3C49	1-17/32
–47	1C47	1-15/32	–47	2C47	1-15/32	–51	3C51	1-19/32
–49	1C49	1-17/32	–49	2C49	1-17/32	–53	3C53	1-21/32
–51	1C51	1-19/32	–51	2C51	1-19/32	–55	3C55	1-23/32
–53	1C53	1-21/32	–53	2C53	1-21/32	–57	3C57	1-25/32
–55	1C55	1-23/32	–55	2C55	1-23/32	–59	3C59	1-27/32

Miscellaneous Nuts

DASH NO.	SIZE
–3	#10-32
–4	1/4-28
–5	5/16-24
–6	3/8-24
–7	7/16-20
–8	1/2-20
–9	9/16-18
–10	5/8-18
–12	3/4-16
–14	7/8-14
–16	1-14

AN310—CASTLE   AN315—PLAIN   AN316—CHECK

AN320—SHEAR   AN360—ENGINE

Steel nuts are cadmium plated per specification QQ-P-416.
Example: AN310-5 is castle nut made of steel and fits a $5/16$ AN bolt.

Cadmium-Plated Steel Locknuts, Nylon Insert

AN364 MS 20364
(THIN)
TO 250°F

CADMIUM PLATED STEEL

AN365 MS 20365
(REGULAR)
TO 250°F

CADMIUM-PLATED STEEL

DASH NO.	SIZE
–440	#4-40
–632	#6-32
–832	#8-32
–1032	#10-32
–428	1/4-28
–524	5/16-24
–624	3/8-24
–720	7/16-20
–820	1/2-20
–918	9/16-18
–1018	5/8-18

Example: MS 20364-624 is self-locking thin steel nut for $3/8$ inch bolt, $3/8$-24 thread.

AN960 – Flat Washer

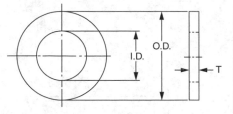

SCREW OR BOLT SIZE	REGULAR			
	DASH NO.	T DIM.	I.D.	O.D.
#2	−2	.032	.099	.250
#3	−3	.032	.105	.250
#4	−4	.032	.125	.312
#6	−6	.032	.149	.375
#8	−8	.032	.174	.375
#10	−10	.063	.203	.438
1/4	−416	.063	.265	.500
5/16	−516	.063	.328	.562
3/8	−616	.063	.390	.625
7/16	−716	.063	.453	.750
1/2	−816	.063	.515	.875
9/16	−916	.063	.578	1.062
5/8	−1016	.063	.640	1.188
3/4	−1216	.090	.765	1.312
7/8	−1416	.090	.890	1.500
1"	−1616	.090	1.015	1.750

AN960　- CADMIUM PLATED CARBON STEEL
AN960A　- ALUMINUM (UNTREATED)
　　　　　(NOT READILY AVAILABLE, USE AN960D)
AN960B　- BRASS
AN960C　- STAINLESS STEEL
AN960D　- ALUMINUM ALLOY, CONDITION T3 OR T4
AN960PD - ALUMINUM ALLOY, ANODIZED

AN380 (MS24665) – Cotter Pin

DIAMETER

LENGTH

CORROSION RESISTING STEEL		DIAMETER & LENGTH	CADMIUM PLATED STEEL	
MS 24665	AN380		MS 24665	AN380
−20	C1-1	1/32 × 3/8	−3	−1-1
−22	C1-2	1/32 × 1/2	−5	−1-2
−24	C1-3	1/32 × 3/4	−7	−1-3
−26	C1-4	1/32 × 1	−9	−1-4
−149	C2-1	1/16 × 3/8	−130	−2-1
151	C2-2	1/16 × 1/2	−132	−2-2
−153	C2-3	1/16 × 3/4	−134	−2-3
−155	C2-4	1/16 × 1	−136	−2-4
−157	C2-5	1/16 × 1-1/4	−138	−2-5
−159	C2-6	1/16 × 1-1/2	−140	−2-6
−161	C2-7	1/16 × 1-3/4	−142	−2-7
−162	C2-8	1/16 × 2	−143	−2-8
−229		5/64 × 3/4		
−231		5/64 × 1		
−298	C3-2	3/32 × 1/2	−281	−3-2
−300	C3-3	3/32 × 3/4	−283	−3-3
−302	C3-4	3/32 × 1	−285	−3-4
−304	C3-5	3/32 × 1-1/4	−287	−3-5
−306	C3-6	3/32 × 1-1/2	−289	−3-6
−308	C3-7	3/32 × 1-3/4	−291	−3-7
−309	C3-8	3/32 × 2	−292	−3-8
−366	C4-2	1/8 × 1/2	−349	−4-2
−368	C4-3	1/8 × 3/4	−351	−4-3
−370	C4-4	1/8 × 1	−353	−4-4
−374	C4-6	1/8 × 1-1/2	−357	−4-6
−377	C4-8	1/8 × 2	−360	−4-8
−379	C4-10	1/8 × 2-1/2	−362	−4-10
		5/32 × 1	−419	−5-4
		5/32 × 1-1/2	−423	−5-6
	C5-9	5/32 × 2-1/4	−427	−5-9

DASH NO.	DIAMETER AND THREAD	COTTER PINS FOR AN310 & AN320
−3	#10-32	AN380-2-1
−4	1/4-28	AN380-2-2
−5	5/16-24	AN380-2-2
−6	3/8-24	AN380-3-3
−7	7/16-20	AN380-3-3
−8	1/2-20	AN380-3-3
−9	9/16-18	AN380-4-4
−10	5/8-18	AN380-4-4
−12	3/4-16	AN380-4-5

Machine Screws

AN500A AND AN501A
MS 35265 AND MS 35266

DRILLED FILLISTER HEAD
SCREWS
CARBON STEEL
CADMIUM PLATED

LENGTH
DIA

AN500A DASH	MS 35265 DASH	THREAD	LENGTH
4-4	–13	#4 – 40	1/4
4-5	–14		5/16
4-6	–15		3/8
4-8	–17		1/2
6-4	–26	#6 – 32	1/4
6-5	–27		5/16
6-6	–28		3/8
6-8	–30		1/2
6-10	–31		5/8
6-12	–32		3/4
8-4	–41	#8 – 32	1/4
8-5	–42		5/16
8-6	–43		3/8
8-8	–45		1/2
8-10	–46		5/8
8-12	–47		3/4
8-14	–48		7/8
8-16	–49		1
10-4	–59	#10 – 24	1/4
10-6	–61		3/8
10-8	–63		1/2
10-10	–64		5/8
10-12	–65		3/4
10-14	–66		7/8
10-16	–67		1
416-8	–79	1/4 – 20	1/2
416-10	–80		5/8

AN501A DASH	MS 35266 DASH	THREAD	LENGTH
10-4	–59	#10 – 32	1/4
10-5	–60		5/16
10-6	–61		3/8
10-8	–63		1/2
10-10	–64		5/8
10-12	–65		3/4
10-16	–67		1
416-10	–80	1/4 – 28	5/8

AN505 MS 35190
AN510 MS 35191
82°
FLAT HEAD

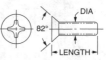

DIA
82°
DIA
LENGTH

AN505 DASH	MS 35190 DASH	THREAD	LENGTH
–4R4	–221	#4 – 40	1/4
–4R6	–223		3/8
–4R8	–225		1/2
–4R10	–227		3/4
–6R4	–234	#6 – 32	1/4
–6R6	–236		3/8
–6R8	–238		1/2
–6R10	–239		5/8
–6R12	–240		3/4
–6R14	–241		7/8
–6R16	–242		1
–8R4		#8 – 32	1/4
–8R6	–251		3/8
–8R8	–253		1/2
–8R10	–254		5/8
–8R12	–255		3/4
–8R16	–257		1
–8R20	–259		1-1/4
–8R24	–261		1-1/2

AN510 DASH	MS 35191 DASH	THREAD	LENGTH
–10R4		#10 – 32	1/4
–10R6			3/8
–10R8			1/2
–10R10			5/8
–10R12			3/4
–10R14			7/8
–10R16			1

Example: AN500A-10-14 (MS 35265-66) is fillister head screw, 10-24 thread and $7/8$ in. long, drilled head.

Example: AN505-8R10 (MS 35191-254) is flat, recessed head, 8-32 thread screw, $5/8$ inch long.

Machine Screws

AN507 MS 24693
100°
FLAT HEAD

CADMIUM-PLATED CARBON STEEL

AN509 MS 24694
100°
FLAT HEAD
STRUCTURAL

AN507 DASH	MS 24693 DASH	THREAD	LENGTH
440R4	S2		1/4
440R5	S3		5/16
440R6	S4		3/8
440R8	S6	#4 – 40	1/2
440R10	S7		5/8
440R12	S8		3/4
440R14	S9		7/8
440R16	S10		1
632R4	S24		1/4
632R5	S25		5/16
632R6	S26		3/8
632R7	S27		7/16
632R8	S28		1/2
632R10	S29	#6 – 32	5/8
632R12	S30		3/4
632R14	S31		7/8
632R16	S32		1
632R20	S34		1-1/4
632R24	S36		1-1/2
832R4	S46		1/4
832R5	S47		5/16
832R6	S48		3/8
832R7	S49		7/16
832R8	S50		1/2
832R10	S51	#8 – 32	5/8
832R12	S52		3/4
832R14	S53		7/8
832R16	S54		1
832R20	S56		1-1/4
832R24	S58		1-1/2
1032R4	S268		1/4
1032R5	S269		5/16
1032R6	S270		3/8
1032R7	S271		7/16
1032R8	S272		1/2
1032R10	S273	#10 – 32	5/8
1032R12	S274		3/4
1032R14	S275		7/8
1032R16	S276		1
1032R20	S278		1-1/4
1032R24	S280		1-1/2

AN509 DASH	MS 24694 DASH	SIZE	GRIP/LENGTH
8R5	S2		.093 - .343
8R6	S3		.093 - .406
8R7	S4		.093 - .468
8R8	S5		.093 - .531
8R9	S6	#8 – 32	.156 - .593
8R10	S7		.218 - .656
8R11	S8		.281 - .718
8R12	S9		.343 - .781
8R13	S10		.406 - .843
8R14	S11		.468 - .906
8R15	S12		.531 - .968
8R16	S13		.593 - 1.031
10R6	S48		.109 - .406
10R7	S49		.109 - .468
10R8	S50		.109 - .531
10R9	S51		.109 - .593
10R10	S52		.187 - .656
10R11	S53		.250 - .718
10R12	S54		.312 - .781
10R13	S55	#10 – 32	.375 - .843
10R14	S56		.437 - .906
10R15	S57		.500 - .968
10R16	S58		.562 - 1.031
10R17	S59		.625 - 1.093
10R18	S60		.687 - 1.156
10R19	S61		.750 - 1.218
10R20	S62		.812 - 1.281
416R7	S94		.140 - .468
416R8	S95		.140 - .531
416R9	S96		.140 - .593
416R10	S97		.140 - .656
416R11	S98		.187 - .718
416R12	S99		.250 - .781
416R13	S100	1/4 – 28	.312 - .843
416R14	S101		.375 - .906
416R15	S102		.437 - .968
416R16	S103		.500 - 1.031
416R17	S104		.562 - 1.093
416R18	S105		.652 - 1.156
416R19	S106		.687 - 1.218
416R20	S107		.750 - 1.281

Example: AN507-832R10 (MS 24693-551) is flat, recessed head, 8-32 thread screw, 5/8 inch long.

Example: AN509-10R16 (MS 24694-558) is flat, recessed head, 10-32 thread, structural screw, nominal length, 1 inch and 9/16 inch nominal grip length.

Machine Screws

AN526 TRUSS HEAD

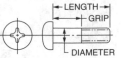

LENGTH

DIAMETER

MS 27039

PAN HEAD
MACHINE SCREW
STRUCTURAL
CROSS RECESSED

LENGTH

GRIP

DIAMETER

DASH NO.	THREAD	LENGTH
632R4		1/4
632R5		5/16
632R6		3/8
632R7		7/16
632R8		1/2
632R9	#6 - 32	9/16
632R10		5/8
632R12		3/4
632R14		7/8
632R16		1
632R18		1 - 1/8
632R20		1 - 1/4
632R24		1 - 1/2
632R28		1 - 3/4
632R32		2
832R4		1/4
832R5		5/16
832R6		3/8
832R7		7/16
832R8		1/2
832R9		9/16
832R10	#8 - 32	5/8
832R12		3/4
832R14		7/8
832R16		1
832R18		1 - 1/8
832R20		1 - 1/4
832R24		1 - 1/2
832R28		1 - 3/4
832R32		2
1032R4		1/4
1032R5		5/16
1032R6		3/8
1032R7		7/16
1032R8		1/2
1032R9		9/16
1032R10	#10 - 32	5/8
1032R12		3/4
1032R14		7/8
1032R16		1
1032R18		1-1/8
1032R20		1-1/4
1032R24		1-1/2
1032R28		1-3/4
1032R32		2

DASH NO.	THREAD	LENGTH	GRIP
0804		.281	.032
0805		.344	.032
0806		.406	.032
0807		.469	.032
0808		.531	.094
0809		.594	.156
0810	#8 - 32	.656	.219
0811		.719	.281
0812		.781	.344
0813		.844	.406
0814		.906	.469
0815		.969	.531
0816		1.031	.594
0818		1.156	.719
0820		1.281	.844
0821		1.344	.906
1-06		.406	.032
1-07		.469	.032
1-08		.531	.062
1-09		.594	.125
1-10		.656	.188
1-11	#10 - 32	.719	.250
1-12		.781	.312
1-13		.844	.375
1-14		.906	.438
1-15		.969	.500
1-16		1.031	.563
4-12	1/4 - 28	.781	.250

Example: AN526-832R10 is truss head, recessed head screw, 8-32 thread and $5/8$ in. long.

Example: MS 27039-0816 is pan-head structural, recessed head, screw 8-32 thread, nominal length 1 in. and nominal grip length of $19/32$ in.

Sheet-Metal Self-Tapping Screws

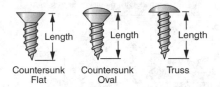

Countersunk Flat Countersunk Oval Truss

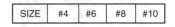

"A" TYPE

"B" TYPE

LENGTH
1/4
3/8
1/2
5/8
3/4
1-
1-1/4
1-1/2

TYPE "A" IS A COARSE-THREADED SCREW WITH A SHARP GIMLET POINT. TYPE "B" OR "Z" (USED WITH TINNERMAN SPEED NUTS) HAS FINER PITCHED THREADS AND IS BLUNT ENDED. A REFINEMENT OF BOTH IS TYPE "AB" THAT HAS THE "B" THREAD AND THE SHARP "A" POINT.

SIZE	#4	#6	#8	#10

Phillips Recessed   Slotted

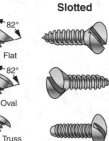

Flat

Oval

Truss

Round

TYPE B AN530

NAS548 (MS 21207)

NAS 548 (MS 21207)
100° PHILLIPS FLAT HEAD
TYPE "B" TAPPING SCREW

NAS 548 (MS 21207)	DIAMETER	#6	#8	#10
	1st DASH NO.	−6	−8	−10

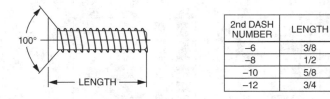

100°

LENGTH

2nd DASH NUMBER	LENGTH
−6	3/8
−8	1/2
−10	5/8
−12	3/4

Example: NAS548-8-8 is #8 Phillips, 100-degree flat-head type B, tapping screw, 1/2 inch long. (NAS548-8-8 is the same as #8 × 1/2, 100-degree flat-head tapping screw.)

Tinnerman Speed Nut
Flat Type

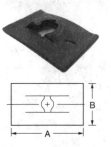

USE WITH TYPE B (BLUNT TAPER AT END) TAPPING SCREWS

PART NUMBER	SCREW SIZE	A LENGTH	B WIDTH
A 1776-4Z-1	4B	.500	.312
A 1181-6Z-1	6B	.515	.312
A 1777-6Z-1	6B	.625	.437
A 1778-8Z-1	8B	.625	.437
A 1779-10Z-1	10B	.875	.500

FOR USE WITH MACHINE SCREWS

PART NUMBER	SCREW SIZE
A 105-440-1	4-40
A 1322-632-1	6-32
A 1322-832-1	8-32

DESIGN VARIATIONS AVAILABLE

A No extrusion on lower leg

B Full extrusion on lower leg

C Straight upper leg

D Corner turned up

E Relief notch

F Corners cut off

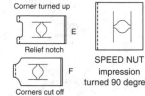

H SPEED NUT impression turned 90 degrees

U Type

SELF-RETAINING "U" TYPE, PRESS EASILY INTO LOCKED-ON POSITION OVER PANEL EDGES OR IN CENTER PANEL LOCATIONS. THEY HOLD THEMSELVES IN A SCREW-RECEIVING POSITION AND ARE IDEALLY SUITED FOR BLIND ASSEMBLY OR HARD-TO-REACH LOCATIONS. IDEALLY SUITED WHERE FULL BEARING SURFACE ON THE LOWER LEG OF THE SPEED NUT IS REQUIRED.

PART NUMBER	SCREW SIZE	DESIGN	PANEL RANGE	LENGTH A	WIDTH B
A 6187-4Z-1	4	CD	.025 - .032	.375	.312
A 1784-6Z-1	6	E	.025 - .051	.640	.437
A 1785-6Z-1	6	E	.025 - .064	.843	.437
A 6052-6Z-1	6	EH	.032 - .040	.468	.500
A 1274-8Z-1	8	DEH	.025 - .032	.500	.500
A 1348-8Z-1	8	AE	.025 - .064	.750	.500
A 1786-8Z-1	8	CEH	.040 - .051	.531	.500
A 1787-8Z-1	8	E	.025 - .064	.843	.437
A 1788-8Z-1	8	AE	.025 - .064	.843	.437
A 1789-8Z-1	8	E	.025 - .051	.640	.437
A 1932-8Z-1	8	BE	.032 - .051	.593	.500
A 1350-10Z-1	10	AE	.025 - .064	.750	.500
A 1758-10Z-1	10	E	.081 - .094	.640	.437
A 1787-10Z-1	10	E	.025 - .064	.843	.437
A 1791-10Z-1	10	EH	.040 - .064	.625	.625
A 1794-10Z-1	10	E	.032 - .064	.640	.437
A 9031-10Z-1	10	EH	.102 - .125	.703	.625

Solid Rivets

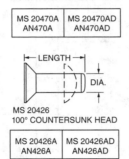

MS 20470 UNIVERSAL HEAD

MS 20470A	MS 20470AD
AN470A	AN470AD

MS 20426
100° COUNTERSUNK HEAD

MS 20426A	MS 20426AD
AN426A	AN426AD

The MS 20426 and MS 20470 types are the most widely used: manufactured to Mil-R-5674. These two types are available in most sized in two materials: "hard" 2117 aluminum alloy (AD) and "soft" 1100 pure aluminum (A).

Example of part no.: MS 20426A3-12 is $3/32$ inch dia., $3/4$ inch long, 100 degrees counter sunk head "soft."

DASH NO.	DIA.	LENGTH
2-6		3/8
2-8	1/16	1/2
2-9		9/16
3-3		3/16
3-4		1/4
3-5	3/32	5/16
3-6		3/8
3-8		1/2
3-12		3/4
4-3		3/16
4-4		1/4
4-5		5/16
4-6		3/8
4-7		7/16
4-8		1/2
4-9	1/8	9/16
4-10		5/8
4-11		11/16
4-12		3/4
4-14		7/8
4-16		1
4-18		1-1/8
4-20		1-1/4
4-30		1-7/8
5-4		1/4
5-5		5/16
5-6		3/8
5-7		7/16
5-8	5/32	1/2
5-9		9/16
5-10		5/8
5-12		3/4
5-16		1
5-22		1-3/8
6-4		1/4
6-5		5/16
6-6		3/8
6-7		7/16
6-8	3/16	1/2
6-10		5/8
6-12		3/4
6-14		7/8

Solid Rivet Identification Chart

Material		Head Marking	AN Material Code	AN425 75° Counter-Sunk Head	AN426 100° Counter-Sunk Head MS 20426*	AN427 100° Counter-Sunk Head MS 20427*	AN430 Round Head MS 20470*	AN435 Round Head MS 20613* MS 20615*	AN441 Flat Head	AN442 Flat Head MS 20470*	AN455 Brazier Head MS 20470*	AN456 Brazier Head MS 20470*	AN470 Universal Head MS 20470*	Heat Treat Before Using	Shear Strength P.S.I.	Bearing Strength P.S.I.
Aluminum Alloy	1100	Plain	A	x	x		x						x	No	10000	25000
	2117T	Recessed Dot	AD	x	x		x			x	x	x	x	No	30000	100000
	2017T	Raised Dot	D	x	x		x			x	x	x	x	Yes	34000	113000
	2017T-HD	Raised Dot	D	x	x		x			x	x	x	x	No	38000	126000
	2024T	Raised Double Dash	DD	x	x		x			x	x	x	x	Yes	41000	136000
	5056T	Raised Cross	B	x	x		x			x	x	x	x	No	27000	90000
	7075-T73	Three Raised Dashes		x	x		x			x	x	x	x	No		
	Carbon Steel	Recessed Triangle				x		x MS 20613*	x					No	35000	90000
	Corrosion Resistant Steel	Recessed Dash	F			x		x MS 20613*						No	65000	90000
	Copper	Plain	C			x		x MS 20615*	x					No	23000	
	Monel	Plain	M			x			x					No	49000	
	Monel (Nickel-Copper Alloy)	Recessed Double Dots	C					x MS 20615*						No	49000	49000
	Brass	Plain						x MS 20615*						No		
	Titanium	Recessed Large and Small Dot			MS 20426				x					No	95000	

x Indicates head shapes and materials available.

*New specifications are for Design purposes

MS Turnbuckles (Clip-Locking)

Clip-Locking Turnbuckles utilize two locking clips instead of lockwire for safetying. The turnbuckle barrel and terminals are slotted lengthwise to accommodate the locking clips. After the proper cable tension is reached the barrel slots are aligned with the terminal slots and the clips are inserted. The curved end of the locking clips expand and latch in the vertical slot in the center of the barrel.

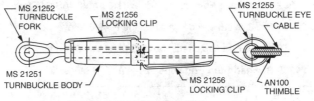

TYPICAL TURNBUCKLE ASSEMBLY

MS Standard Drawings for clip-locking turnbuckles supersede various AN Drawings for conventional (lockwire type) turnbuckle parts and NAS Drawings for clip-locking turnbuckle parts. Refer to the following cross reference tables for AN and NAS equivalents.

MS 21251 Turnbuckle Barrel

Supersedes AN155 and NAS649 barrels. MS 21251 items can replace AN155 items of like material and thread, but the AN155 items cannot replace the MS 21251 items. MS 21251 items are interchangeable with the NAS 649 items of like material and thread. MS 21251 barrels are available in brass (QQ-B-637, composition 2 or MIL-T-6945), steel (cadmium plated to QQ-P-416, type 2, class 3) or aluminum alloy (anodized to MIL-A-8725). The cross reference table shows equivalent items made of brass.

MS 21251 DASH NO.	ROPE DIA.	THREAD SIZE	AN155 DASH NO	NAS 649 DASH NO.	USES MS 21256 CLIP DASH NO.
B2S	1/16	6–40	B8S	B8S	−1
B2L	1/16	6–40	B8L	B8L	−2
B3S	3/32	10–32	B16S	B16S	−1
B3L	3/32	10–32	B16L	B16L	−2
B5S	5/32	1/4–28	B32S	B32S	−1
B5L	5/32	1/4–28	B32L	B32L	−2
B6S	3/16	5/16–24	B46S	B46S	−1
B6L	3/16	5/16–24	B46L	B46L	−2
B8L	1/4	3/8–24	B80L	B80L	−2
B9L	9/32	7/16–20	B125L	B125L	−3
B10L	5/16	1/2–20	B175L	B175L	−3

Terminals

MS items can replace AN items of like thread except for the -22 and -61 sizes, but the AN items cannot replace the MS items. MS items are interchangeable with the NAS items of like thread except for the -22 and -61 sizes. These MS terminals are available only in steel cadmium plated to QQ-P-416, type 2, class 3. Available with right-hand (R) or left-hand (L) threads.

MS 21252 Turnbuckle Fork supersedes AN161 and NAS 645 forks.

MS 21254 Pin Eye supersedes AN165 and NAS 648 eyes.

MS 21255 Cable Eye supersedes AN170 and NAS 647 eyes.

MS 21260 Swaged Stud End supersedes AN669 studs.

MS 21252 MS 21254 MS 21255 DASH NOS.		WIRE ROPE DIA.	THREAD SIZE	AN161 AN165 AN170 DASH NOS.		NAS 645 NAS 648 NAS 647 DASH NOS.	
RH THD	LH THD			RH THD	LH THD	RH THD	LH THD
-2RS	-2LS	1/16	6-40	-8RS	-8LS	-8RS	-8LS
-2RL*	-2LL*	1/16	6-40	—	—	—	—
-3RS	-3LS	3/32	10-32	-16RS	-16LS	-16RS	-16LS
-3RL	-3LL	3/32	10-32	-16RL	-16LL	-16RL	-16LL
-5RS	-5LS	5/32	1/4-28	-32RS	-32LS	-32RS	-32LS
-5RL	-5LL	5/32	1/4-28	-32RL	-32LL	-32RL	-32LL
-6RS	-6LS	3/16	5/16-24	-46RS	-46LS	-46RS	-46LS
-6RL	-6LL	3/16	5/16-24	-46RL	-46LL	-46RL	-46LL
-8RL	-8LL	1/4	3/8-24	-80RL	-80LL	-80RL	-80LL
-9RL	-9LL	9/32	7/16-20	-125RL	-125LL	-125RL	-125LL
-10RL	-10LL	5/16	1/2-20	-175RL	-175LL	-175RL	-175LL

*MS 21254 and MS 21255 eyes only; MS 21252 fork not made in this size.

MS 21256 Turnbuckle Clip

Made of corrosion resistant steel wire, QQ-W-423, composition FS302, condition B. These are NOT interchangeable with the NAS 651 clips. Available in 3 sizes: MS 21256-1, -2, and -3. For applications, see the MS 21251 Turnbuckle Barrel Cross Reference Chart.

MS 21260 Swaged Stud End

These clip-locking terminals are available in corrosion resistant steel and in cadmium-plated carbon steel. MS 21260 items can replace AN669 items of the same dash numbers, but the AN669 items cannot always replace the MS 21260 items.

Example: The AN "equivalent" (the AN equivalent would not be clip-locking) for MS 21260 L3RH would be AN669-L3RH. There would be no AN equivalent for a MS 21260 FL3RH, since AN669 terminals are not available in carbon steel.

PART NUMBER	THREAD	CABLE DIA.	DESCRIPTION
MS 21255-3LS	10-32	3/32	Eye End (for cable)
-3RS	10-32	3/32	
MS 21256-1	–	–	Clip (for short barrels)
-2	–	–	Clip (for long barrels)
MS 21260-S2LH	6-40	1/16	
-S2RH	6-40	1/16	
-S3LH	10-32	3/32	
-S3RH	10-32	3/32	
-L3LH	10-32	3/32	End (for cable)
-L3RH	10-32	3/32	
-S4LH	1/4-28	1/8	
-S4RH	1/4-28	1/8	
-L4LH	1/4-28	1/8	
-L4RH	1/4-28	1/8	

PART NUMBER	THREAD	CABLE DIA.	DESCRIPTION
MS 21251-B2S	6-40	1/16	
-B3S	10-32	3/32	
-B3L	10-32	3/32	Barrel (Body), Brass
-B5S	1/4-28	5/32	
-B5L	1/4-28	5/32	
MS 21252-3LS	10-32	3/32	
-3RS	10-32	3/32	Fork (Clevis End)
-5RS	1/4-28	5/32	
MS 21254-2RS	6-40	1/16	
-3LS	10-32	3/32	
-3RS	10-32	3/32	Eye End (for pin)
-5LS	1/4-28	5/32	
-5RS	1/4-28	5/32	

AN Turnbuckle Assemblies

AN 130 ASSEMBLY

|◄———— LENGTH ————►|

CABLE EYE   BARREL AN 155   FORK AN 161
AN 170

AN 140 ASSEMBLY

|◄———— LENGTH ————►|

CABLE EYE   BARREL AN 155   CABLE EYE
AN 170   AN 170

AN 135 ASSEMBLY

|◄———— LENGTH ————►|

CABLE EYE   BARREL AN 155   PIN EYE
AN 170   AN 165

AN 150 ASSEMBLY

|◄———— LENGTH ————►|

FORK AN 161   BARREL AN 155   FORK AN 161

Turnbuckles consist of a brass barrel, and two steel ends, one having a right-handed thread and the other a left-handed thread. Types of turnbuckle ends are cable eye, pin eye, and fork. Turnbuckles illustrated on this page show four recommended assemblies. Turnbuckle barrels are made of brass: cable eyes, pin eyes, and forks of cadmium-plated steel.

Example: AN155-8S (Barrel; length 2¼ inch) AN161-16RS (Fork; short, R.H. thread)

DASH NO.	LENGTH		STRENGTH	THREAD SIZE
	AN130	OTHERS		
8S	4-1/2	4-1/2	800	6-40
16S	4-1/2	4-1/2	1600	10-32
16L	8	8	1600	10-32
22S	4-17/32	4-1/2	2200	1/4-28
22L	8-1/32	8	2200	1/4-28
32S	4-19/32	4-1/2	3200	1/4-28
32L	8-7/64	8	3200	1/4-28

Swaging Terminals

AN663C   MS 20663

BALL AND DOUBLE SHANK

DASH NUMBER	CABLE DIAMETER	FINISHED DIAMETER	
		BALL	SHANK
2	1/16	.190	.112
3	3/32	.250	.140
4	1/8	.312	.190
5	5/32	.375	.218
6	3/16	.437	.250

AN664C   MS 20664

BALL AND SHANK

DASH NUMBER	CABLE DIAMETER	FINISHED DIAMETER	
		BALL	SHANK
2	1/16	.190	.112
3	3/32	.250	.140
4	1/8	.312	.190
5	5/32	.375	.218
6	3/16	.437	.250

AN666   MS 21259

STUD END

DASH NUMBER		CABLE DIAMETER	THREAD SIZE
RIGHT HAND	LEFT HAND		
2RH	2LH	1/16	#6-40
3RH	3LH	3/32	#6-40
4RH	4LH	1/8	1/4-28
5RH	5LH	5/32	1/4-28

AN667   MS 20667

FORK END

DASH NUMBER	CABLE DIAMETER	PIN HOLE DIAMETER	SLOT WIDTH
2	1/16	.190	.093
3	3/32	.190	.108
4	1/8	.190	.195
5	5/32	.250	.202

AN668   MS 20668

EYE END

DASH NUMBER	CABLE DIAMETER	PIN HOLE DIAMETER	BLADE WIDTH
2	1/16	.190	.088
3	3/32	.190	.103
4	1/8	.190	.190
5	5/32	.250	.197

(FOR SAFETY WIRE)   AN669   STUD END   MS 21260 (SLOTTED FOR CLIP)

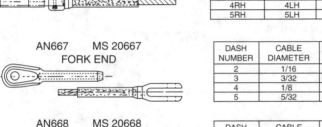

TURNBUCKLE END – SHORT			
DASH NUMBER		CABLE DIAMETER	THREAD SIZE
RIGHT HAND	LEFT HAND		
S2RH	S2LH	1/16	6-40
S3RH	S3LH	3/32	10-32
S4RH	S4LH	1/8	1/4-28

TURNBUCKLE END – LONG			
DASH NUMBER		CABLE DIAMETER	THREAD SIZE
RIGHT HAND	LEFT HAND		
L2RH	L2LH	1/16	6-40
L3RH	L3LH	3/32	10-32
L4RH	L4LH	1/8	1/4-28

Plumbing Fittings AN774-AN932

Material:

Aluminum alloy...........................(code D)

Steel..(code, absence of letter)

Brass...(code B)

Aluminum bronze........................(code Z—for AN819 sleeve)

Size: The dash number following the AN number indicates the size of the tubing (or hose) for which the fitting is made, in 16ths of an inch. This size measures the O.D. of tubing and the I.D. of hose. Fittings having pipe threads are coded by a dash number, indicating the pipe size in 8ths of an inch. The material code letter, as noted above, follows the dash number.

Example: AN822-5-4D is an aluminum 90° elbow for 5/6 inch tubing and 1/4 inch pipe thread.

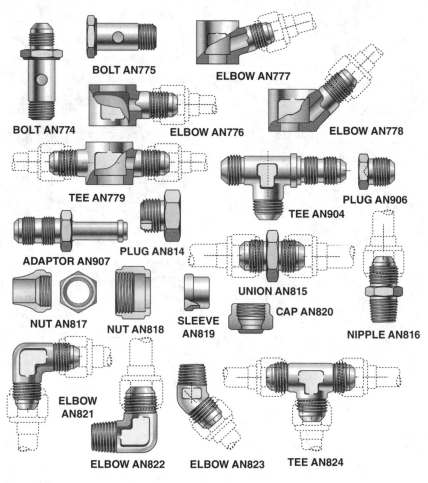

BOLT AN775

ELBOW AN777

BOLT AN774

ELBOW AN776

ELBOW AN778

TEE AN779

PLUG AN906

TEE AN904

PLUG AN814

ADAPTOR AN907

UNION AN815

NUT AN817

NUT AN818

SLEEVE AN819

CAP AN820

NIPPLE AN816

ELBOW AN821

ELBOW AN822

ELBOW AN823

TEE AN824

Plumbing Fittings (Continued)

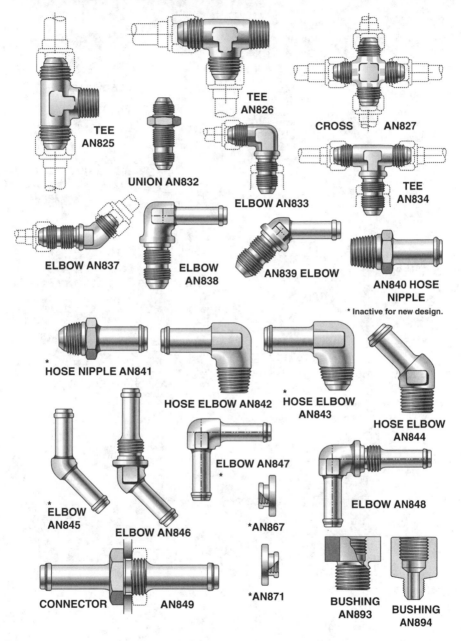

TEE
AN825

TEE
AN826

CROSS   AN827

UNION AN832

ELBOW AN833

TEE
AN834

ELBOW AN837

ELBOW
AN838

AN839 ELBOW

AN840 HOSE
NIPPLE

* Inactive for new design.

* HOSE NIPPLE AN841

HOSE ELBOW AN842

* HOSE ELBOW
AN843

HOSE ELBOW
AN844

* ELBOW
AN845

ELBOW AN846

ELBOW AN847
*

*AN867

ELBOW AN848

CONNECTOR   AN849

*AN871

BUSHING
AN893

BUSHING
AN894

Plumbing Fittings (Continued)

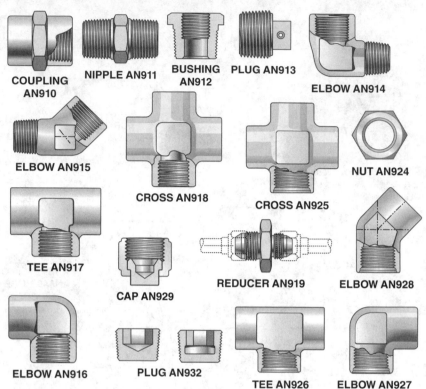

COUPLING AN910

NIPPLE AN911

BUSHING AN912

PLUG AN913

ELBOW AN914

ELBOW AN915

CROSS AN918

CROSS AN925

NUT AN924

TEE AN917

CAP AN929

REDUCER AN919

ELBOW AN928

ELBOW AN916

PLUG AN932

TEE AN926

ELBOW AN927

Additional Standard
Parts (Patented)

The following pages illustrate a few fastener types widely used on high-performance aircraft. These fasteners are designed and manufactured by various companies, are patented, and are generally known by their trade names.

It is emphasized that the following pages are in no way a complete list of patented fasteners available. Representative examples only are shown for illustrative purposes. All of these fasteners require special installation tools and procedures. Installation manuals are available from the manufacturers.

CONVERSION TABLE

NAS NUMBERS TO CHERRY RIVET NUMBERS

(A COMPLETE CONVERSION TABLE OF CHERRY RIVET NUMBERS IS AVAILABLE UPON REQUEST)

BULBED CHERRYMAX® RIVETS

HEAD STYLE	NAS NUMBER	CHERRY NUMBER	RIVET MATERIAL	STEM MATERIAL
UNIVERSAL HEAD	NAS 1738B 1738E 1738M 1738MW 1738C 1738CW	CR 2249 2239 2539 2539P 2839 2839CW	5056 Aluminum 5056 Aluminum Monel Monel, Cad. Plt'd. Inconel 600 Inconel 600, Cad. Plt'd.	Alloy Steel, Cad. Plt'd. Inconel 600 Inconel 600 Inconel 600 A 286 CRES A 286 CRES
COUNTERSUNK HEAD (MS 20426)	NAS 1739B 1739E 1739M 1739MW 1739C 1739CW	CR 2248 2238 2538 2538P 2838 2838CW	5056 Aluminum 5056 Aluminum Monel Monel, Cad. Plt'd. Inconel 600 Inconel 600, Cad. Plt'd.	Alloy Steel, Cad. Plt'd. Inconel 600 Inconel 600 Inconel 600 A 286 CRES A 286 CRES
UNISINK HEAD	— — — —	CR 2235 2245 2545 2845	5056 Aluminum 5056 Aluminum Monel Inconel 600	Inconel 600 Alloy Steel, Cad. Plt'd. Inconel 600 A 286 CRES
COUNTERSUNK HEAD (156°)	— —	CR 2540 2840	Monel Inconel 600	Inconel 600 A 286 CRES

BULBED CHERRYMAX® RIVETS
NAS 1738 UNIVERSAL HEAD

PROCUREMENT SPECIFICATION NAS 1740 IS APPLICABLE TO NAS 1736 RIVETS.

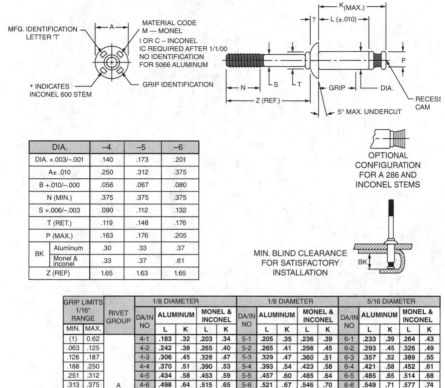

DIA.	–4	–5	–6
DIA. +.003/–.001	.140	.173	.201
A± .010	.250	.312	.375
B +.010/–.000	.056	.067	.080
N (MIN.)	.375	.375	.375
S +.006/–.003	.090	.112	.132
T (RET.)	.119	.148	.176
P (MAX.)	.163	.176	.205
BK Aluminum	.30	.33	.37
BK Monel & inconel	.33	.37	.61
Z (REF)	1.65	1.63	1.65

OPTIONAL
CONFIGURATION
FOR A 286 AND
INCONEL STEMS

MIN. BLIND CLEARANCE
FOR SATISFACTORY
INSTALLATION

GRIP LIMITS 1/16" RANGE		RIVET GROUP	1/8 DIAMETER					1/8 DIAMETER					5/16 DIAMETER				
			DA/IN NO	ALUMINUM		MONEL & INCONEL		DA/IN NO	ALUMINUM		MONEL & INCONEL		DA/IN NO	ALUMINUM		MONEL & INCONEL	
MIN.	MAX.			L	K	L	K		L	K	L	K		L	K	L	K
(1)	0.62		4-1	.183	.32	.203	.34	5-1	.205	.35	.236	.39	6-1	.233	.39	.264	.43
.063	.125		4-2	.242	.38	.265	.40	5-2	.265	.41	.298	.45	6-2	.293	.45	.326	.49
.126	.187		4-3	.306	.45	.328	.47	5-3	.329	.47	.360	.51	6-3	.357	.52	.389	.55
.188	.250		4-4	.370	.51	.390	.53	5-4	.393	.54	.423	.58	6-4	.421	.58	.452	.61
.251	.312		4-5	.434	.58	.453	.59	5-5	.457	.60	.485	.64	6-5	.485	.65	.514	.68
.313	.375	A	4-6	.498	.64	.515	.65	5-6	.521	.67	.546	.70	6-6	.549	.71	.577	.74
.376	.437		4-7	.562	.71	.578	.72	5-7	.585	.73	.610	.76	6-7	.613	.78	.639	.80
.438	.500		4-8	.626	.77	.640	.78	5-8	.649	.80	.673	.83	6-8	.677	.84	.702	.88
.501	.562		4-9	.690	.84	.703	.84	5-9	.713	.86	.735	.89	6-9	.741	.91	.764	.93
.563	.625							5-10	.777	.93	.798	.95	6-10	.806	.97	.827	.99
.626	.687							5-11	.841	.99	.860	1.01	6-11	.869	1.04	.889	1.05
.688	.750												6-12	.933	1.10	.952	1.11

	(1)
RIVET DIA.	MIN. GRIP
1/8	.020
5/32	.025
3/16	.030

RIVET GROUP REFERS TO SHIFT-POINT SETTING OF RIVETER.

hi-shear RIVET IDENTIFICATION CHART

PART NUMBER	IDENTIFICATION (HI-SHEAR CORPORATION) HEAD MARKING OR COLOR CODE	MATERIAL	PHYSICAL PROPERTIES ROOM TEMP.	HEAD TYPE	TOLERANCES MINIMUM C'SX HEAT HEIGHT	TOLERANCES MINIMUM SHANK DIAMETER	HI-SHEAR COLLAR TO ORDER	SUGGESTED MAXIMUM TEMP. FOR USE	CHARACTERISTICS	NAS OR CUSTOMER PART NUMBER
NAS177	(+)	Alloy Steel	125,000 – 150,000 psi Tensile	Csk		.0025	NAS179		Inactive. See HS47, HS47 PB and HS48. or NAS1054 and NAS1055.	
NAS178				Flat						
NAS179	Grey	2117–T4 Aluminum Alloy							Inactive. See HS15 or NAS528.	
NAS528	Red	2024–T4 Aluminum Alloy							Hi-Shear Collar used in combination with 160,000 – 180,000 psi tensile Hi-Shear Rivet Pins. Same as HS15.	
NAS1806 thru NAS1816	NAS 1810 HV	6A1–4V Titanium Alloy	95,000 psi Shear Minimum	Flat		.0005	HS15		Chamfered lead style. Used in high performance aircraft where weight, shank, and hole tolerances are critical.	
NAS1906 Thru NAS1916	NAS 1910 HV	6A1–4V Titanium Alloy	95,000 psi Shear Minimum	Csk		.0005	HS15		Chamfered lead style. Used in high performance aircraft where weight, shank, and hole tolerances are critical.	
HS10	(⊙)	Alloy Steel	160,000 – 180,000 psi Tensile	Csk		.0025	HS15		Stud Rivet Pin–fastens primary structure and provides for a means of attaching removable elements.	
HS15	Red	2024–T4 Aluminum Alloy						200°F	Hi-Shear Collar used in combination with 160,000 – 180,000 psi tensile Hi-Shear Rivet Pins.	NAS528
HS23 HS23A	H	7075–T6 Aluminum Alloy		Csk	.002	.001	HS24		Hi-Shear 100° head. Higher shear and tension allowables than DD Rivets. Small 'head permits countersinking in thin materials. "A" signifies sodium dichromate seal. For oversize, use HS39P or HS41P.	
HS24	Blue	2117–T4 Aluminum Alloy						200°F	Hi-Shear Collar used in combination with 7075-T6 aluminum alloy Hi-Shear Rivet Pins.	
HS25 HS25A	H	7075–T6 Aluminum Alloy		Csk	.002	.001	HS24		MS 20426 style head. Higher shear and tension allowables than DD Rivets. "A" signifies sodium dichromate seal. For oversize, use HS25,32 or HS25,64.	

PART NUMBER	IDENTIFICATION (HI-SHEAR CORPORATION) HEAD MARKING OR COLOR CODE	MATERIAL	PHYSICAL PROPERTIES ROOM TEMP.	HEAD TYPE	TOLERANCES MINIMUM C'SX HEAD HEIGHT	TOLERANCES MINIMUM SHANK DIAMETER	HI-SHEAR COLLAR TO ORDER	SUGGESTED MAXIMUM TEMP. FOR USE	CHARACTERISTICS	NAS OR CUSTOMER PART NUMBER
HS25.32 HS25.64	No Head Marking 1/32 – Red 1/64 – Blue	7075–T6 Aluminum Alloy		Csk	.002	.001	HS24		Oversizes for HS25.	
HS26 HS26A	(H)	7075–T6 Aluminum Alloy		Flat		.001	HS24		Higher shear and tension allowable than DD Rivets. "A" signifies sodium dichromate seal. For oversize, use HS26.32 or HS26.64.	
HS26.32 HS26.64	No Head Marking 1/32 – Red 1/64 – Blue	7075-T6 Aluminum Alloy		Flat		.001	HS24		Oversizes for HS26.	
HS30	No Marking	Alloy Steel	160,000 – 180,000 psi Tensile	No Head		.0025	HS15 HS24 HS32		Dowel Pin grooved for collar on both ends. Lighter and stronger than Taper Pin. Precision fit molds to irregular surfaces (both sides). HS30 ground after plating; HS30 P plated after grind.	
HS32	Silver (Cadmium Plate)	Low Carbon Steel							Hi-Shear Collar used in combination with 160,000 – 180,000 psi tensile Hi-Shear Rivets Ferrous material applications.	
HS39P HS40P	(.64)	Alloy Steel	160,000 – 180,000 psi Tensile	Csk Flat	.002	.0011	HS15		1/64 oversize Rivet Pin for HS23, HS26, HS47, HS48, HS51 P, and HS52 P, HS39 PB and HS40 PB – Type II plating.	
HS41P HS42P	(.32)	Alloy Steel	160,000 – 180,000 psi Tensile	Csk Flat	.002	.0011	HS15 or HS46		1/32 oversize Rivet Pin for HS23, HS26, HS47, HS48, HS51 P, and HS52 P, HS41 PB and HS42 PB – Type II plating.	
HS47 HS48	(H)	Alloy Steel	160,000 – 180,000 psi Tensile	Csk Flat	.002	.0025	HS15		Chamfered lead style used in design where shank commercial tolerances are acceptable. For oversize use HS39 P, HS40 P, HS41 P, or HS42 P, HS47 PB – Type II plating.	NAS1055 NAS1054
HS51P HS52P	(H)	Alloy Steel	160,000 – 180,000 psi Tensile	Csk Flat	.002	.0011	HS15		Chamfered lead style - plated after grind. Used in design where shank and hole tolerances are critical. For oversize use HS39 P, HS40 P, HS41 P, or HS42 P, HS51 PB – Type II plating.	NAS525 NAS529
HS53	Red	2024-T6 Aluminum Alloy							Countersank flanged Hi-Shear Collar used in double dimple applications. Used in combination with 160,000 – 180,000 psi tensile Hi-Shear Rivet Pins.	

PART NUMBER	IDENTIFICATION (HI-SHEAR CORPORATION) HEAD MARKING OR COLOR CODE	MATERIAL	PHYSICAL PROPERTIES ROOM TEMP.	HEAD TYPE	TOLERANCES MINIMUM		HI-SHEAR COLLAR TO ORDER	SUGGESTED MAXIMUM TEMP. FOR USE	CHARACTERISTICS	NAS OR CUSTOMER PART NUMBER
					C'SK HEAD HEIGHT	SHANK DIAMETER				
HS54	Blue	2117–T4 Aluminum Alloy							Countersunk flanged Hi-Shear Collar used in double dimple applications. Used in combination with 7075–T6 aluminum alloy Hi-Shear Rivet Pins.	
HS60	Natural	321 Stainless Steel						1600°F	Hi-Shear Collar for use in high temperature applications. Maximum temperature governed by conditions of application or use.	
HS60M	Black	"F" Monel or 400 Monel	R_B 96 (Max.)					1000°F	Hi-Shear Collar used in high temperature applications to 900°F.	
HS61	61	Type 431 Stainless Steel	125,000 psi Shear Minimum	Csk	.002	.0005	HS60 or HS60M	450°F	Used in high strength or temperature applications where shank and hole tolerances are critical. For oversize, use HS139, HS140, HS141, or HS142.	
HS62	62			Flat						
HS65	65	Type 305 Stainless Steel	125,000 psi Tensile	Csk	.002	.0025	HS60 or HS60M •		*Use HS60 Collars for non-magnetic applications. HS60 M for other applications.	
HS66	66			Flat						
HS67	67	Type 431 Stainless Steel	125,000 psi Shear Minimum	Csk	.002	.0025	HS60 or HS60M	450°F	Used in high strength or temperature applications where shank commercial tolerances are acceptable. For oversize, use HS139, HS140, HS141, or HS142.	
HS68	68			Flat						
HS90	Natural	A-286 High Temp. Alloy						1200°F	Hi-Shear Collar used in non-magnetic and high temperature applications.	
HS91	91	A-286 High Temp. Alloy	95,000 psi Shear Minimum	Csk	.002	.0005	HS60M or HS90	1200°F (See Note)	Used in strength and temperature applications where shank and hole tolerances are critical. NOTE: For use in non-magnetic applications.	
HS92	92			Flat						
HS104	Natural	Inconel 600 per AMS5665						1500°F	Used in combination with HS131 and HS132 Pins. For use at high temperature applications.	
HS106	No Head Marking	Alloy Steel	160,000 – 180,000 psi Tensile	Flat		Knurled Shank	HS24		Rivet Pin with threaded stud. Fastens primary structure and provides threaded stud to attach removal items.	

PART NUMBER	IDENTIFICATION (HI-SHEAR CORPORATION) HEAD MARKING OR COLOR CODE	MATERIAL	PHYSICAL PROPERTIES ROOM TEMP.	HEAD TYPE	TOLERANCES MINIMUM		HI-SHEAR COLLAR TO ORDER	SUGGESTED MAXIMUM TEMP. FOR USE	CHARACTERISTICS	NAS OR CUSTOMER PART NUMBER
					C'SX HEAD HEIGHT	SHANK DIAMETER				
HS108	• HS	Alloy Steel	160,000 180,000 psi Tensile	Protruding		.0025	HS15		Stud Rivet Pin Fastens primary structure and provides stud to attach removable items.	
HS131	131	Inconel X–750	160,000 psi Tensile	Csk	.002	.0025	HS104	1500°F	Used at high temperature applications where shank commercial tolerances are acceptable	
HS132	132			Flat						
HS139	139	Type 431 Stainless Steel	125,000 psi Shear Minimum	Csk	.002	.0005	HS60M	450°F	1/64 oversize for HS61, HS62, HS67, and HS68.	
HS140	140			Flat						
HS141	141	Type 431 Stainless Steel	125,000 psi Shear Minimum	Csk	.002	.0005	HS60M	450°F	1/32 oversize for HS61, HS62, HS67, and HS68.	
HS142	142			Flat						
HS149	H 149	6 – 4 T$_1$ Alloy	160,000 – 180,000 psi Tensile	Csk	.002	.0005	HS15 HS167 HS234		Used in high performance aircraft where weight material fatigue, shank and hole tolerances are critical.	NAS1806 NAS1906
HS150	H 150			Flat						
HS154	H 154 P	Type H-11 Steel per AMS6485	156,000 psi Shear Min. R$_C$ 50–55	Flat	.002	.0025	HS90		Used in high temperature applications. "P" code – cadmium plate. "N" code – diffused nickel.	
HS155	155 P			Csk						
HS159	H 159	A-286 High Temp. Alloy	160,000 – 180,000 psi Tensile	Csk	.002	.001	HS60M		1/64 oversize for HS91 and HS92.	
HS160	H 160			Flat						
HS161	H 161	A-286 High Temp. Alloy	160,000 – 180,000 psi Tensile	Csk	.002	.0015	HS60M		1/32 oversize for HS91 and HS92.	
HS162	H 162			Flat						
HS167	Natural	Ti–50A Com. Pure Titanium						700°F	Hi-Shear Collar used in combination with HS149 and HS150 Pins and in non-magnetic applications.	
HS234	Violet	2219–T6 Aluminum Alloy						425°F	Hi-Shear Collar used at elevated temperatures.	

TRI-WING®

1. NAS STANDARDS AND SPECIFICATIONS
2. AIRLINE AND MANUFACTURERS APPROVAL
3. THREE-WING RECESSED DESIGN PERMITS EASY IDENTIFICATION
4. REDUCED WORK EFFORT BY THE OPERATOR RESULTS FROM LESS END THRUST
5. CLOSE-TOLERANCE CONTROL OF THE RECESS AND THE DRIVER BIT ACHIEVE OPTIMUM PERFORMANCE
6. IMPROVER DRIVER BIT LIFE
7. PART NUMBERS ARE STAMPED ON THE FASTENER HEADS
8. DRIVER NUMBERS ARE STAMPED ON THE FASTENER HEADS
9. DRIVER BITS ARE NUMBERED WITH RECESS SIZE TO ELIMINATE MISMATCH PROBLEMS
10. POWER DRIVER OPERATIONS OF THE TRI-WING INSURE POSITIVE ENGAGEMENT-REDUCING THE CHANCE OF SURFACE DAMAGE TO THE ADJACENT STRUCTURE

TRI-WING® STANDARDS

NAS NUMBER	DIAMETER	HEAD STYLE	THREAD TYPE	MATERIAL	CLASSIFICATION
NAS 4104-4116 NAS 4204-4216 NAS 4304-4316	250 – 1,000"	100°	Long	Alloy Steel Cres Titanium	Bolt
NAS 4400-4416 NAS 4500-4516 NAS 4600-4616	112 – 1,000"	100°	Short	Alloy Steel Cres Titanium	Bolt
NAS 4703-4716 NAS 4803-4816 NAS 4903-4916	190 – 1,000"	100° Reduced	Short	Alloy Steel Cres Titanium	Bolt
NAS 5000-5006 NAS 5100-5106 NAS 5200-5206	112 – 375"	Pan	Short	Alloy Steel Cres Titanium	Screw
NAS 5300-5306 NAS 5400-5406 NAS 5500-5506	112 – 375"	Fillister	Full	Alloy Steel Cres Titanium	Screw
NAS 5600-5606 NAS 5700-5706 NAS 5800-5806	112 – 375"	100°	Full	Alloy Steel Cres Titanium	Screw
NAS 5900-5903 NAS 6000-6003 NAS 6100-6103	112 – 190"	Hex	Full	Alloy Steel Cres Titanium	Screw

APPLICABLE SPECIFICATIONS

TRI-WING recess specification – NAS 4000   TRI-WING driver specification – NAS 4001
Alloy Steel process specification – NAS 4002   Cres process specification – NAS 4003
Titanium process specification – NAS 4004

TRI-WING® is a registered trademark of PHILLIPS SCREW COMPANY

STANDARDS COMMITTEE
FOR HI-LOK® PRODUCTS
2600 SKYPARK DRIVE, TORRANCE, CALIFORNIA 90509

HI-SHEAR CORPORATION (Patent Holder) – Federal Code Ident. No. 73197
VOI-SHAN DIV, VSI CORP. (Licensee) – Federal Code Ident. No. 92215
STANDARD PRESSED STEEL CO. (Licensee) – Federal Code Ident. No. 56878

HI-LOK PIN IDENTIFICATION CHART

Issued: 3-20-68
Revised: October 1972

HI-LOK PIN PART NO.	PIN HEAD STYLE / APPLICATION	MATERIAL	HEAT TREAT	SHANK DIA. TOL.	SUGGESTED MAXIMUM TEMP. FOR USE	GRIP VARIATION	RECOMMENDED COMPANION HI-LOK COLLARS	NEXT OVERSIZE	CHARACTERISTICS
HL10	Protruding / Shear	6A1-4V Titanium	95,000 psi Shear Minimum	.0005 or .0010	750°F or Sub. to Finish	1/16"	HL70 HL94 HL79 HL97 HL82 HL379	HL110	Used when weight conservation is essential and where pin shank and hole tolerances are critical. Anti-galling finish available for use with all types of Hi-Lok collar materials.
HL11	100° Flush / Shear	6A1-4V Titanium	95,000 psi Shear Minimum	.0005 or .0010	750°F or Sub. to Finish	1/16"	HL70 HL94 HL79 HL97 HL82 HL379	HL111	Used when weight conservation is essential and where pin shank and hole tolerances are critical. Anti-galling finish available for use with all types of Hi-Lok collar materials.
HL12	Protruding / Tension	6A1-4V Titanium	160,000 psi Tensile Minimum	.0005 or .0010	750°F or Sub. to Finish	1/16"	HL75 HL198 HL86 HL280	HL112	Used when weight conservation is essential and where pin shank and hole tolerances are critical. Anti-galling finish available for use with all types of Hi-Lok collar materials.
HL13	100° Flush MS 24694 / Tension	6A1-4V Titanium	160,000 psi Tensile Minimum	.0005 or .0010	750°F or Sub. to Finish	1/16"	HL75 HL198 HL86 HL280	HL113	Used when weight conservation is essential and where pin shank and hole tolerances are critical. Anti-galling finish available for use with all types of Hi-Lok collar materials.
HL14	Protruding / Shear	H-11 Steel Alloy	156,000 psi Shear Minimum	.001	900°F or Sub. to Finish	1/16"	HL288 HL574	HL214	Used in high temperature applications where pin shank and hole close tolerances are required.
HL15	100° Flush / Shear	H-11 Steel Alloy	156,000 psi Shear Minimum	.001	900°F or Sub. to Finish	1/16"	HL288 HL574	HL215	Used in high temperature applications where pin shank and hole close tolerances are required.
HL16	Protruding / Tension	H-11 Steel Alloy	260,000-280,000 psi Tensile	.001	900°F or Sub. to Finish	1/16"	HL89 HL273	HL216	Used in high temperature applications where pin shank and hole close tolerances are required.

EXAMPLES ONLY – CATALOG HAS PIN TYPES THRU HL 1317

hi-Lok® hi-tigue® PRODUCTS
2600 SKYPARK DRIVE, TORRANCE, CALIFORNIA 90509

STANDARDS
MANUAL

HI-LOK® HI-TIGUE® PIN IDENTIFICATION CHART

HI-LOK HI-TIIGE PIN PART NO.	PIN HEAD STYLE APPLICATION	MATERIAL	HEAT TREAT	SHANK DIA. TOL.	SUGGESTED MAXIMUM TEMP. FOR USE	GRIP VARIATION	RECOMMENDED COMPANION HI-LOK HI-TIGUE COLLARS		NEXT OVERSIZE	CHARACTERISTICS
HLT10	Protruding Shear	6A1–4 V Titanium Alloy	95,000 psi Shear Minimum	.001	600°	1/16"	HLT70 HLT71	HLT94 HLT97	HLT 110	Used where weight conservation and high fatigue life is critical. Pins are designed for easy installation in interference fit holes. Anti-galling finish available for use with all types of Hi-Lok Hi-Tigue collar materials.
HLT11	100° Flush Crown Shear	6A1–4 V Titanium Alloy	95,000 psi Shear Minimum	.001	600°	1/16"	HLT70 HLT71	HLT94 HLT97	HLT111	Used where weight conservation and high fatigue life is critical. Pins are designed for easy installation in interference fit holes. Anti-galling finish available for use with all types of Hi-Lok Hi-Tigue collar materials.
HLT12	Protruding Tension	6A1–4 V Titanium Alloy	160,000 psi Tensile Minimum	.001	600°	1/16"	HLT78 HLT87		HLT112	Used where weight conservation and high fatigue life is critical. Pins are designed for easy installation in interference fit holes. Anti-galling finish available for use with all types of Hi-Lok Hi-Tigue collar materials.
HLT13	100° Flush MS 24694 Tension	6A1–4 V Titanium Alloy	160,000 psi Tensile Minimum	.001	600°	1/16"	HLT78 HLT87		HLT113	Used where weight conservation and high fatigue life is critical. Pins are designed for easy installation in interference fit holes. Anti-galling finish available for use with all types of Hi-Lok Hi-Tigue collar materials.
HLT18	Protruding Shear	Alloy Steel	95,000 psi Shear Minimum	.001	450°	1/16"	HLT70 HLT71	HLT94 HLT97	HLT118	Pins are designed for easy installation in interference fit holes.
HLT19	100° Flush Crown Shear	Alloy Steel	95,000 psi Shear Minimum	.001	450°	1/16"	HLT70 HLT71	HLT94 HLT97	HLT119	Pins are designed for easy installation in interference fit holes.
HLT22	Protruding Shear	6A1–6V –2Sn Titanium Alloy	108,000 psi Shear Minimum	.001	600°	1/16"	HLT70 HLT71	HLT94 HLT97	HLT122	Same as HLT10 except for material and heat treat.

EXAMPLES ONLY – CATALOG HAS PIN TYPES THRU HLT 931

STANDARDS COMMITTEE
FOR HI-LOK® PRODUCTS
2600 SKYPARK DRIVE, TORRANCE, CALIFORNIA 90509

HI-SHEAR CORPORATION (Patent Holder) – Federal Code Ident. No. 73197
VOI-SHAN DIV, VSI CORP. (Licensee) – Federal Code Ident. No. 92215
STANDARD PRESSED STEEL CO. (Licensee) – Federal Code Ident. No. 56878

HI-LOK COLLAR IDENTIFICATION CHART

Issue Date: November 1972

HI-LOK COLLAR PART NO.	COLLAR MATERIAL	COLLAR FINISH COLOR OR PLATING	WASHER MATERIAL	WASHER FINISH COLOR OR PLATING	SUGGESTED MAXIMUM TEMP. FOR USE	GRIP VARIATION	APPLICATION	NEXT OVERSIZE	CHARACTERISTICS
HL70	2024-T6 Aluminum Alloy	See Drawing	2024 or 5052 Aluminum Alloy	Blue or Grey	300°F	1/16"	Shear	HL79 or HL80	For use with shear head pins except those made of aluminum alloy. Optional washer.
HL75	303 Series Stainless Steel	See Drawing	17-4 PH, 17-7 PH or PH15-7 Mo Stainless Steel	See Drawing	300°F or Sub. to Finish	1/16"	Tension	HL375	Self-aligning collar assembly. For use on sloped surfaces up to 7° maximum. Fits standard and 1/64" oversize tension head pins.
HL77	2024-T6 Aluminum Alloy	See Drawing	2024 or 5052 Aluminum Alloy	Black	300°F	1/16"	Shear	HL377	For use with aluminum alloy pins in shear applications. Optional washer.
HL78	A-286 Hi-Temp. Alloy	See Drawing	300 Series Stainless Steel	See Drawing	1200°F or Sub. to Finish	1/16"	Tension	HL278	Used in high temperature applications. Anti-galling lubricant available for use in titanium pins. Optional washer.
HL79	2024-T6 Aluminum Alloy	Red	N/A	N/A	300°F	1/16"	Shear	HL84	For standard and 1/64" oversize for HL70. For use with Hi-Lok Automatic Feed Driver Tools and shear head pins except those made of aluminum alloy.
HL82	2024-T6 Aluminum Alloy	See Drawing	17-4 PH, 17-7 PH or PH15-7 Mo Stainless Steel	See Drawing	300°F	1/16"	Shear	HL382	Self-aligning collar assembly. For use on sloped surfaces up to 7° maximum. Fits standard and 1/64" oversize pins. Use with shear head pins except those made of aluminum alloy.
HL87	303 Series Stainless Steel	See Drawing	300 Series Stainless Steel	Cadmium Plate	700°F or Sub. to Finish	1/16"	Tension	HL93	For standard and 1/64" oversize for HL86. Optional washer.

EXAMPLES ONLY – CATALOG HAS COLLAR TYPES THRU HL 1775

Appendix

Tap Drill Sizes—American (National) Screw Thread Series

NATIONAL COARSE THREAD SERIES MEDIUM FIT. CLASS 3 (NC)					NATIONAL FINE THREAD SERIES MEDIUM FIT. CLASS 3 (NF)				
Size and Threads	Dai. of body for thread	Body Drill	Tap Drill		Size and threads	Dai. of body for thread	Body Drill	Tap Drill	
			Pref'd dia. of hole	Nearest stand'd Drill Size				Pref'd dia. of hole	Nearest stand'd Drill Size
					0-80	.060	52	.0472	3/34
1-64	.073	47	.0575	#53	1-72	.073	47	.0591	#53
2-56	.086	42	.0682	#51	2-64	.086	42	.0700	#50
3-48	.099	37	.078	5/64	3-56	.099	37	.0810	#46
4-40	.112	31	.0866	#44	4-48	.112	31	.0911	#42
5-40	.125	29	.0995	#39	5-44	.125	25	.1024	#38
6-32	.138	27	.1063	#36	6-40	.138	27	.113	#33
8-32	.164	18	.1324	#29	8-36	.164	18	.136	#29
10-24	.190	10	.1472	#26	10-32	.190	10	.159	#21
12-24	.216	2	.1732	#17	12-28	.216	2	.180	#15
1/4-20	.250	1/4	.1990	#8	1/4-28	.250	F	.213	#3
5/16-18	.3125	5/16	.2559	#F	5/16-24	.3125	5/16	.2703	I
3/8-16	.375	3/8	.3110	#5/16"	3/8-24	.375	3/8	.332	Q
7/16-14	.4375	7/16	.3642	U	7/16-20	.4375	7/16	.386	W
1/2-13	.500	1/2	.4219	27/64"	1/2-20	.500	1/2	.449	7/16"
9/16-12	.5625	9/16	.4776	31/64"	9/16-18	.5625	9/16	.506	1/2"
5/8-11	.625	5/8	.5315	17/32"	5/8-18	.625	5/8	.568	9/16"
3/4-10	.750	3/4	.6480	41/64"	3/4-16	.750	3/4	.6688	11/16"
7/8-9	.875	7/8	.7307	49/64"	7/8-14	.875	7/8	.7822	51/64"
1-8	1.000	1.0	.8376	7/8"	1-14	1.000	1.0	.9072	49/84"

Standard AN aircraft bolts are threaded in National Fine, Class 3 (NF) thread series.

Wire and Sheet Metal Gage Table

American or Brown & Sharpe for Aluminum & Brass Sheet	Gauge	U.S. Standard Gauge for Steel & Plate Iron & Steel
.3648	00	.3437
.3249	0	.3125
.2893	1	.2812
.2576	2	.2656
.2294	3	.2391
.2043	4	.2242
.1819	5	.2092
.1620	6	.1943
.1443	7	.1793
.1285	8	.1644
.1144	9	.1495
.1019	10	.1345
.0907	11	.1196
.0808	12	.1046
.0720	13	.0897
.0641	14	.0747
.0571	15	.0673
.0508	16	.0598
.0453	17	.0538
.0403	18	.0478
.0359	19	.0418
.0320	20	.0359
.0285	21	.0329
.0253	22	.0299
.0226	23	.0269
.0201	24	.0239
.0179	25	.0209
.0159	26	.0179

Ultimate and Shear Strength of Typical Aluminum Alloys

Alloy and temper	Ultimate strength, psi	Shearing strength, psi
1100-0	13,000	9,500
1100-H14	17,000	11,000
1100-H18	24,000	13,000
3003-0	16,000	11,000
3003-H14	21,000	14,000
3003-H18	29,000	16,000
2017-T4	62,000	36,000
2117-T4	43,000	26,000
2024-0	26,000	18,000
2024-T4	68,000	41,000
2024-T36	70,000	42,000
Alclad		
2024-T3	62,000	40,000
Alclad		
2024-T36	66,000	41,000
5052-0	29,000	18,000
5052-H14	37,000	21,000
5052-H18	41,000	24,000
6061-0	18,000	12,500
6061-T6	45,000	30,000
7075-0	33,000	22,000
7075-T6	82,000	49,000
Alclad		
7075-0	32,000	22,000
7075-T6	76,000	46,000

Chemical Flashpoints for Various Liquids Used in the Aircraft Industry

A liquid's flashpoint is the lowest temperature at which it will give off enough flammable vapor at or near its surface in mixture with air and a spark or flame so that it ignites. If the flashpoint, expressed as a temperature in degrees, is lower than the temperature of the ambient air, the vapors will ignite readily in air with a source of ignition. Those of higher temperature are relatively safer.

Chemical Flashpoints in Degrees Fahrenheit and Celsius

	Flashpoint (°F)	Flashpoint (°C)
Acetone	0	−17.8
Alcohol (Denatured)	60	15.6
Alcohol (Ethyl)	55	12.8
Alcohol (Methyl, Methonol, Wood)	54	12.2
Alcohol (Isopropyl)	53	11.7
Benzine (Petroleum Ether)	<50	< −46.0
Benzol (Benzene)	12	−11.1
Diluent A	−10	−23.0
Ethyl Acetate	24	−4.4
Ethyl Ether	−49	−45.0
Fuel, Jet A	95−145	35.0−62.8
Fuel, JP-4	−10	−23.0
Fuel, JP-5	95−145	35.0−62.8
Fuel Oil	100	37.8
Gasoline	−45	−42.8
Kerosene	100−165	+37.8−+73.9
K.U.L.	120	48.9
Lacquer	0−80	−17.8−+26.7
Lacquer Thinner	40	4.4
Methyl Cellosolve Acetate (MCA)	132	55.6
Methyl Ethyl Ketone (MEK)	30	−1.1
Methyl Iso-Butyl Ketone (MIBK)	73	22.8
Mineral Spirits (Turpentine Subst.)	85	29.4
Naphtha VM&P	20−45	−6.7−+7.2
Naphtha, Petroleum Ether	<0	< −17.8
Paint, Liquid	0−80	−17.8−+26.7
Shell 40	140	60.0
Stoddard Solvent	100−110	37.8−43.3
Styrene	90	32.2
Thinner (Wash)	20	−6.7
Toluol (Toluene)	40	4.4
Turpentine	95	35.0
Varnish	10−80	−12.2−+26.7
Xylene	63	17.2

Glossary

Alclad Trademark used by the Aluminum Company of America to identify a group of high-strength, sheet-aluminum alloys covered with a high-purity aluminum.

Allowance An intentional difference permitted between the maximum material limits of mating parts. It is the minimum clearance (positive allowance) or maximum interference (negative allowance) between parts.

Alloy A substance composed of two or more metals or of a metal and a nonmetal intimately united, usually by being fused together and dissolving in each other when molten. All rivets and sheet metal used for structural purposes in aircraft are alloys.

A-N (or AN) An abbreviation for Air Force and Navy; especially associated with Air Force and Navy standards or codes for materials and supplies. Formerly known as *Army-Navy standards*.

AN specifications Dimensional standards for aircraft fasteners developed by the Aeronautical Standards Group.

Angle of head In countersunk heads, the included angles of the conical underportion or bearing surface, usually 100 degrees.

Bearing surface Supporting or locating surface of a fastener with respect to the part to which it fastens (mates). Loading of a fastener is usually through the bearing surface.

Blind riveting The process of attaching rivets where only one side of the work is accessible.

Broaching The process of removing metal by pushing or pulling a cutting tool, called a *broach*, along the surface.

Bucking To brace or hold a piece of metal against the opposite side of material being riveted to flatten the end of rivet against material.

Bucking bars A piece of metal held by a rivet bucker against the opposite end of a rivet being inserted into material.

Burnishing The process of producing a smooth surface by rubbing or rolling a tool over the surface.

Burr   A small amount of material extending out from the edge of a hole, shoulder, etc. Removal of burrs is called *burring* or *deburring*.

Center punch   A hand punch consisting of a short steel bar with a hardened conical point at one end, which is used to mark the center of holes to be drilled.

Chip   A small fragment of metal removed from a surface by cutting with a tool.

Chip chaser   A flat, hooked piece of metal inserted between materials being drilled to remove chips.

Chord   The straight line that joins the leading and trailing edges of an airfoil.

Coin dimpling   A form of countersinking resulting from squeezing a single sheet of material between a male and female die to form a depression in the material and allow the fastener head to be flush with the material's surface.

Collar   A raised ring or flange of material placed on the head or shank of a fastener to act as a locking device.

Corrosion   The wearing away or alteration of a metal or alloy either by direct chemical attack or electrochemical reaction.

Counterboring   The process of enlarging for part of its depth a previously formed hole to provide a shoulder at bottom of the enlarged hole. Special tools, called *counterbores*, are generally used for this operation.

Countersinking   The process of beveling or flaring the end of a hole. Holes in which countersunk head-type fasteners are to be used must be countersunk to provide a mating bearing surface.

CRT   Cathode-ray tube.

Cryogenic temperatures   Extremely cold or very low temperatures that are associated with ordinary gases in a liquid state.

Defect   A discontinuity that interferes with the usefulness of the part, a fault in any material, or a part detrimental to its serviceability.

Die   One of a pair of hardened metal blocks used to form, impress, or cut out a desired shape; a tool for cutting external threads.

Drilling   The process of forming holes by means of specialized, pointed cutting tools, called *drills*.

Edge distance   The distance from the centerline of the rivet hole to the nearest edge of the sheet.

Elongated Stretched out, lengthened, or long in proportion to width.

Endurance limit The maximum stress that a fastener can withstand without failure for a specified number of stress cycles (also called *fatigue limit*).

Facing A machining operation on the end, flat face, or shoulder of a fastener.

Fastener A mechanical device used to hold two or more bodies in definite positions, with respect to each other.

Faying surface The side of a piece of material that contacts another piece of material being joined to it.

Ferrous metal A metal containing iron. Steel is a ferrous metal. See *nonferrous metal*.

Fit A general term used to signify the range of tightness that results from the application of a specific combination of allowances and tolerances in the design of mating parts.

Grinding The process of removing a portion of the surface of a material by the cutting action of a bonded abrasive wheel.

Grip In general, the grip of a fastener is the thickness of the material or parts that the fastener is designed to secure when assembled.

Gun As used in riveting, some form of manual or powered tool used to drive and fasten rivets in place.

Hole finder A tool used to exactly locate and mark the position of holes to be drilled to match the location of a pilot or predrilled hole.

Increment One of a series of regular consecutive additions; for example, $\frac{1}{8}$, $\frac{2}{8}$, and $\frac{3}{8}$.

Interference fit A thread fit having limits of size so prescribed that an interference always results when mating parts are assembled.

Jig A device that holds and locates a piece of work and guides the tools that operate upon it.

Lightening holes Holes that are cut in material to lessen the overall weight of the material, but not weaken the structural strength.

Nondestructive testing (NDT) An inspection or examination of the aircraft for defects on the surface or inside of the material, or hidden by other structures, without damaging the part. Sometimes called *nondestructive inspection (NDI)*.

Nonferrous metal Metal that does not contain iron. Aluminum is a nonferrous metal.

Peen   To draw, bend, or flatten by hammering; the head of a hammer opposite the striking face.

Pilot hole   A small hole used for marking or aligning to drill a larger hole.

Pin   A straight cylindrical or tapered fastener, with or without a head, designed to perform a semipermanent attaching or locating function.

Pitch distance   The distance measured between the centers of two adjoining rivets.

Plating   The application of a metallic deposit on the surface of a fastener by electrolysis, impact, or other suitable means.

Puller   A device used to form or draw certain types of rivets.

Punching   The process of removing or trimming material through the use of a die in a press.

Quick-disconnect   A device to couple (attach) an air hose to air-driven equipment that can be rapidly detached from the equipment.

Ream   To finish a drilled or punched hole very accurately with a rotating fluted tool of the required diameter.

Reference dimension   A reference dimension on a fastener is a dimension without tolerance used for information purposes only.

Rivet   A short, metal, boltlike fastener, without threads, which is driven into place with some form of manual or powered tool.

Rivet set   A small tool (generally round), having one end shaped to fit a specific-shaped rivet head, that fits in a rivet gun to drive the manufactured head of the rivet.

Scribe   A pointed steel instrument used to make fine lines on metal or other materials.

Sealant   A compound or substance used to close or seal openings in a material.

Shaving   A cutting operation in which thin layers of material are removed from the surfaces of the product.

Shear strength   The stress required to produce a fracture when impressed vertically upon the cross section of a material.

Shim   A thin piece of sheet metal used to adjust space.

Shoulder   The enlarged portion of the body of a threaded fastener or the shank of an unthreaded fastener.

Skin, structural   A sheathing or coating of metal placed over a framework to provide a covering material.

Sleeve   A hollow, tubular part designed to fit over another part.

Soft   The condition of a fastener that has been left in the as-fabricated temper, although made from a material that can be, and normally is, hardened by heat treatment.

Spot-face   To finish a round spot on a rough surface, usually around a drilled hole, to provide a good seat to a rivet head.

Standard fastener   A fastener that conforms in all respects to recognized standards or specifications.

Substructure   The underlying or supporting part of a fabrication.

Swaging   Using a swage tool to shape metal to a desired form.

Torque   A turning or twisting force that produces or tends to produce rotation or torsion.

Tolerance   The total permissible variation of a size. Tolerance is the difference between the limits of size.

Upsetting   The process of increasing the cross-section area of a rivet, both longitudinally and radially, when the rivet is driven into place.

Index

Note: Page numbers referencing figures are followed by an *"f"*.